THE FRAGILE ENVIRONMENT

Today we cannot afford to ignore the impact of the human species on its environment. In this topical volume, eight distinguished international authorities identify and analyse areas of rapid and serious change, and outline approaches to some of the outstanding questions. Andrew Goudie's opening chapter places the present environmental crises in chronological perspective, exploring the changing impact of human populations on the environment since prehistoric times, with particular reference to forest cover, soil erosion and animal extinctions. Other contributions then focus on animal species, and the likely effect of human activities on the atmosphere and global climatic patterns. Finally, a former NASA scientist puts these and other environmental problems into a new perspective, as the global environment is examined from space.

The book originates in the highly successful second series of Darwin College Lectures, delivered in Cambridge under the title 'Man and the Environment'.

THE FRAGILE ENVIRONMENT

The Darwin College Lectures

Edited by
Laurie Friday
and
Ronald Laskey

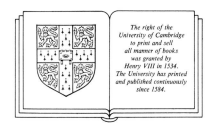

The right of the
University of Cambridge
to print and sell
all manner of books
was granted by
Henry VIII in 1534.
The University has printed
and published continuously
since 1584.

CAMBRIDGE UNIVERSITY PRESS
Cambridge
New York   Port Chester   Melbourne   Sydney

Published by the Press Syndicate of the University of Cambridge
The Pitt Building, Trumpington Street, Cambridge CB2 1RP
40 West 20th Street, New York, NY 10011, USA
10 Stamford Road, Oakleigh, Melbourne 3166, Australia

First published 1989
First paperback edition 1991

Printed in Great Britain at the University Press, Cambridge

British Library Cataloguing in Publication Data
The Fragile Environment. – (Darwin College lectures).
1. Evironment. Conservation
I. Friday, L.E.   II. Laskey, R.A.
III. Darwin College
333.7'2

Library of Congress Cataloguing in Publication Data
The Fragile Environment/edited by L.E. Friday and R.A. Laskey.
p.   cm. "Darwin College lectures."
Includes index.
1. Ecology.
2. Nature conservation.
3. Conservation of natural resources.
I. Friday, L.E. (Laurie Elizabeth), 1955–
II. Laskey, R.A.
III. Title: Darwin College lecture series.
QH541.F695 1989
333.7'2–dc19   88–28551   CIP

ISBN 0 521 36337 3   hardback
ISBN 0 521 42266 3   paperback

Contents

Contributors

Professor Bert Bolin,
Department of Meteorology,
University of Stockholm

Professor Partha Dasgupta,
Faculty of Economics and Politics,
University of Cambridge

Dr Marian Stamp Dawkins,
Department of Zoology,
University of Oxford

Professor Andrew Goudie,
Department of Geography,
University of Oxford

Professor Robert May,
Department of Biology,
Princeton University

Dr Norman Myers,
Headington,
Oxford

Dr Gordon Wells,
Lunar and Planetary Institute,
Houston

Dr Roger Whitehead,
The Dunn Nutrition Unit,
Cambridge

Editorial note

The Darwin College Lecture Series was inaugurated in 1986 to provide a range of public lectures on topics of general interest. The first series was entitled *Origins* and it considered subjects such as the origin of the universe, the origin of man, and the origin of language. The present volume contains the second series which was delivered in Cambridge in 1987, under the original title 'Man and the Environment'. We are grateful to Richard Grove, Christopher Viney and Jamie Whyte for their help in organising this second series.

The contributors to each series have been selected from a wide range of disciplines to allow a broad look at the topic. Although the contributors are acknowledged specialists in their fields, the lectures are aimed at non-specialist readers.

Preface

All living organisms interact with their environments. They are influenced by a host of environmental factors and, to some extent, they modify their environment. The human species is no exception. We respond to patterns of temperature, rainfall, abundance of food and other resources and to the incidence of competitors, predators and disease. However, the relationship of mankind with the environment is unique. The extent and degree to which quite small populations can modify the environment is unparalleled, and this combines with the success of the species to make an overwhelming impact.

Although the environment has some intrinsic resilience to deleterious change, the rate of growth of the human population is such that its environmental effects are far outstripping the recovery potential of the earth. As the population continues to grow at an ever-increasing rate, so the capacity of the earth to support life is being eroded: reports of famine, deforestation, loss of plant and animal species, soil erosion and atmospheric pollution are becoming all too familiar.

However, for the first time in history we are in a position to begin to evaluate our many and complex relationships with our environment and to attempt to tackle the deleterious consequences of our activities. Major new initiatives among all disciplines concerned with the environment are beginning to get under way, with the aim of understanding the workings of the biosphere, and particularly man's effects upon it.

Many of the problems associated with overexploitation of the environment are multinational in origin and are consequently very difficult to resolve. Before drastic changes in public policy can be introduced, a sound base of scientific facts is required. The gathering of the necessary information and the implementation of changes must be considered an urgent priority for mankind. International cooperation and commitment on an unprecedented scale will be necessary to achieve any degree of success.

This book does not contain a recipe for tackling the problems of protecting the world environment from further degradation. Rather, it identifies and analyses some areas of rapid and serious change and it outlines approaches to some of the outstanding fundamental questions concerning the environment.

Inevitably, these represent only a proportion of the vast array of pressing issues, but the eight contributors come from a wide range of backgrounds and each contributes an individual perspective.

In the first chapter, Andrew Goudie sets the present environmental crises in a historical perspective. Drawing together threads from archaeology, geography and palaeoecology he describes the nature and scale of man's impact through forest clearance, hunting, industry and urbanisation in prehistoric times.

In considering the future of forests, Norman Myers underlines the diversity and untapped resources of tropical moist forests, and the vital contribution of all types of forest to the water and carbon balance of the globe. The present rate of destruction of both tropical and temperate forests is alarming, and it is in mankind's own interests to rise to the challenge of arresting it.

Animals form a very important part of our environment. Marian Dawkins examines our attitudes to animals and exposes the paradox of our ambivalent feelings and actions towards other species. In the next chapter, Robert May demonstrates the great uncertainty about exactly how many species we share the planet with. He draws attention to the kinds of information which would provide the means to answer this question, and examines the predictive value of models of the relative abundance of species.

The pressing problem of chronic food shortages is discussed by Roger Whitehead, drawing on experience from nutritional studies in The Gambia. He shows that the vast majority of African nations are poised on a fine dividing line between survival and famine, for economic as well as environmental reasons.

The problem of exhaustible mineral resources is considered from an economist's viewpoint by Partha Dasgupta. Phosphates, some trace elements, and especially fossil fuels, are likely to run out in the forseeable future; the solution must lie in far-sighted research and the development of new technology. In the next chapter, Bert Bolin shows that the earth's climate has been changing in cyclic ways for millennia. With this in mind, he goes on to assess the effects on global temperature patterns of the ever-accelerating change in carbon dioxide and ozone levels in the atmosphere due to man's activities.

The final chapter by Gordon Wells, who at one time trained astronauts in earth observation at NASA, draws together many of the threads running through the book. He demonstrates the value of photographs of the earth taken from space which supply data, unobtainable by other means, on changing land use, climate, marine pollution and volcanic activity.

In all, the global environment consists of a multitude of living and non-living components which interact in an immensely complex way. Changes in just one component can have consequences for many others, and mankind is causing sweeping changes at an ever-increasing rate. The chapters which follow analyse individual aspects of some of these changes and examine their consequences for our Fragile Environment.

[1]

The changing human impact

Andrew Goudie

Introduction

Much of our knowledge of the history of human relationships with the environment has resulted from close, long-standing links between the disciplines of geography and archaeology. Traditionally, such studies have centred on a desire to explain in terms of environmental influences either the distribution of human activities in space or the changing fortunes of human groups through time. For example, Huntington and his co-workers interpreted migration and the decline of cities in Central Asia in terms of environmental changes and saw dry phases as the driving force of the *Pulse of Asia* (1907). The archaeologist Gordon Childe saw the adoption of domestication and cultivation as a revolutionary response to desiccation in post-glacial times. More recently, many of the great collaborative ventures in environmental archaeology have sought to place archaeological sites in their environmental context. A new discipline – Geoarchaeology – has emerged.

In this chapter, however, I wish to approach the subject from a different perspective, seeking to outline some of the ways in which prehistoric groups have caused environmental changes. I shall concentrate on the use of fire, hunting, agriculture and settlement, with particular reference to the post-glacial peoples of Britain, and also on the exploitation of minerals. My hypothesis is that the changes brought about by these activities were rather substantial, though I recognise that the impacts will have varied from area to area, according to length of human occupation, technological level and population numbers.

Human life probably first made its appearance on our planet some three or so million years ago, before the start of the Ice Age (see the timescale in the Appendix to this chapter). The oldest remains have been found in sediments from the Rift Valleys of East Africa and, though less securely dated, from the dolomite cave fills in the Transvaal in South Africa. Likewise, the oldest records of human activity and technology – pebble tools of Oldowan type (crude stone tools which consist of a pebble with one end chipped into a rough cutting edge) have been found with human bone remains in the same areas. For example, at Lake Turkana in northern Kenya, and the Omo Valley in southern Ethiopia, tool-bearing beds of volcanic material have been dated using isotopic means at about 2.6 million years old, while another bed at Olduvai Gorge in Tanzania, 'The Grand Canyon of human evolution', has been dated by the Leakeys and their co-workers using similar methods at 1.75 million years old. At the same time the first larger brained hominid (*Homo habilis*) appeared at Olduvai and Koobi Fora. Using a sharp stone flake, hominids were able to cut through the thick tough hide of a large carcass and obtain quantities of meat unavailable even to the large carnivores until the carcass had sufficiently decomposed. Groups of hominids could have driven other predators from kills. Microscopic examination of Oldowan tools from Koobi Fora (northern Kenya) has revealed that, in addition to being used for cutting meat, they were also used for cutting grass stems or reeds and for scraping and sawing wood. More sophisticated bifacial stone artefacts (handaxes and cleavers) of the Acheulean tradition first appeared in East Africa around 1.6 and 1.5 million years ago. Many of the Acheulean artefacts at Olorgesaillie (Kenya) were brought there from some distance away and, such was the scale of this transport, it seems highly likely that, by this stage in their development, people had invented some sort of bag for carrying their possessions from one place to another. Thus far, however, the human impact on the environment must have been relatively slight even in East Africa. Population levels were undoubtedly low, groups were probably small and highly scattered, and large areas of the Old World had yet to be entered.

As Table 1.1 shows, the duration of human influence has varied in different areas depending on the time it took for diffusion to occur from the African heartland. The dates given are necessarily tentative, approximate and controversial due to the imperfections of the fossil record, the limitations of human bone preservation, and the inadequacies of dating techniques.

Table 1.1. *Dates of human arrivals*

	Source	Date (years BP)*
AFRICA	Klein (1983)	2 700 000–2 900 000
EUROPE	Champion *et al.* (1984)	*c.* 1.6 m.y. but most post-350 000
BRITAIN	Green *et al.* (1981)	*c.* 200 000
JAPAN	Ikawa-Smith (1980)	*c.* 50 000
NEW GUINEA	Bulmer (1982)	*c.* 50 000
AUSTRALIA	Allen *et al.* (1977)	*c.* 40 000
N. AMERICA	Irving (1985)	15 000–40 000
S. AMERICA	Guidon & Delibrias (1986)	32 000
IRELAND	Edwards (1985)	9000
CARIBBEAN	Morgan & Woods (1986)	4500
POLYNESIA	Kirch (1982)	2000
MADAGASCAR	Battistini & Verin (1972)	*c.* 500 AD
NEW ZEALAND	Green (1975)	700–800 AD

* See the timescale on page 19.

Sources for Table 1.1:

Klein, R.G. (1983). The stone age prehistory of southern Africa. *Annual Review of Anthropology*, **12**, 25–48.

Champion, T., Gramble, C., Shennan, S. & Whittle, A. (1984), *Prehistoric Europe*. London: Academic Press.

Green, H.S. *et al.* (1981). Pontnewydd Cave in Wales – a new Middle Pleistocene hominid site. *Nature*, **294**, 707–13.

Ikawa-Smith, F. (1980). Current issues in Japanese archaeology. *American Scientist*, **68**, 134–45.

Bulmer, S. (1982). Human ecology and cultural variation in prehistoric New Guinea. *Monographiae biologicae*, **42**, 169–206.

Allen, J., Golson, J. and Jones, R. (eds.) (1977). *Sunda and Sahul. Prehistoric Studies in Southeast Asia, Melanesia and Australia*. London: Academic Press.

Irving, W.M. (1985). Context and chronology of early man in the Americas. *Annual Review of Anthropology*, **14**, 529–55.

Guidon, N. and Delibrias, G. (1986). Carbon 14 dates point to man in The Americas 32 000 years ago. *Nature*, **321**, 769–71.

Edwards, K.J. (1985). The anthropogenic factor in vegetational history. In K.J. Edwards & W.P. Warren (eds.) *The Quaternary History of Ireland*. London: Academic Press, pp. 187–200.

Morgan, G.S. & Woods, C.A. (1986). Extinction and the zoogeography of West Indian land mammals. *Biological Journal of the Linnean Society*, **28**, 167–203.

Kirch, P.V. (1982) Advances in Polynesian prehistory: Three decades in review. *Advances in World Archaeology*, **2**, 52–102.

Battistini, R. & Verin, P. (1972). Man and the environment in Madagascar. *Monographiae biologicae*, **21**, 311–337.

Green, R.C. (1975). Adaptation and change in Maori culture. *Monographiae biologicae*, **27**, 591–661.

Fire

At some point, man acquired one of the most potent means at his disposal for environmental alteration – fire. Gowlett *et al.* have claimed the discovery of evidence for deliberate manipulation of fire at Chesowanja in Kenya from over 1.4 million years ago. This evidence is controversial, but there is good evidence for the early use of fire in temperate areas. Fires at certain Chinese sites, for example, have been dated at 0.7–1.0 million years ago.

The importance of fire was recognised over a hundred years ago by that father of the conservation movement, George Perkins Marsh (1864, p. 119):

> The destruction of the woods, then, was man's first physical conquest, his first violation of the harmonies of inanimate nature. Primitive man had little occasion to fell trees for fuel, or for the construction of dwellings, boats, and the implements of his rude agriculture and handicrafts. Windfalls would furnish a thin population with a sufficient supply of such material . . . the accidental escape and spread of fire, or, possibly, the combustion of forests by lightning, must have

first suggested the advantages to be derived from the removal
of too abundant and extensive woods, and, at the same time,
have pointed out the means by which a large tract of surface
could readily be cleared of much of this natural incumbrance.

There are many reasons why prehistoric man may have used fire
extensively. The burning of forest increases supportable population sizes for
herbivorous game species: nutrients are released, sunlight access permits
quicker growth of plants available to ground-living herbivores, vigorous
sprouting may increase the amount of food available to browsing animals,
and the germination of seeds may be stimulated. There is evidence that burnt
areas may have ungulate population sizes 300–700% greater than those found
in unburnt areas (Fig. 1.1).

Furthermore, terrain opened up by fire would increase the mobility of
hunting groups and make it easier to see and catch beasts. Fire also gave to
man, for the most part a diurnal creature, security by night from other
predators; it allowed him to move to colder areas; it allowed him to break up
stone to facilitate stone tool manufacture, and the fireside, as Sauer (1969)
pointed out, was the beginning of social living, the place of communication
and reflection. Fire may also have improved human health by reducing the
danger of illness from parasitic and disease organisms that often con-
taminated raw foods (e.g. parasitic worms in raw meat) and destroying the

Fig. 1.1. The burning of sour, highveld grassland in Swaziland,
southern Africa, is practised on a regular basis to encourage the
sprouting of fresh shoots of grass, so that animals, whether wild or
domesticated, receive improved fodder. The relative treelessness of
the highveld may in part be caused by the frequency of burning.

microorganisms that were often responsible for food spoilage. Fire could also be employed to dry surplus plant foods and meat so that they could be preserved for long periods.

Subsequently, fire became a potent force for clearing land for agriculture and for improving grazing land for domestic animals. It was much used by such diverse groups as the cattle-keepers of Africa, the hunter–gatherers of South America from Amazonia to Tierra del Fuego (the Land of Fire) and the aboriginals of Australia. Even in areas with low densities of hunter and gatherer populations (e.g. Australia and Kalahari) the effects of fire are clear.

Although people have used fires for all the reasons mentioned above, before one can assess the impact of man's fires, one must realise how low in importance fires started by human action are in comparison with those caused naturally, especially by lightning which strikes the land surface of the globe 100 000 times each day. Other natural fires may result from such processes as spontaneous combustion or from sparks produced by falling boulders, landslides and volcanic activity.

Given this important caveat, there is evidence that fires set by man have played an important role in the formation of various major types of vegetation that were formerly thought to be the mature, natural vegetation in a particular climate. This applies, for instance, to some savannas, mid-latitude grasslands, heathlands and shrublands (such as are shown in Fig. 1.2 or the

Fig. 1.2. Among the major vegetation types that may owe much of their character to the effects of long-continued burning are the shrubby heathlands of some areas with Mediterranean climate. This example comes from just north of Perth, western Australia.

garrigue of the Mediterranean lands and the chaparral of the south-west United States). These vegetation types reflect some general ecological consequences of fire: assisted seed germination, stimulated vegetative reproduction of many woody and herbaceous species, release of mineral nutrients, and opening up of habitats, thereby promoting species diversity. Fire may also create soil erosion and slope instability. As Pyne (1982, p. 3) has written:

> Hardly any plant community in the temperate zone has escaped fire's selective action, and thanks to the radiation of Homo sapiens throughout the world, fire has been introduced to nearly every landscape on earth. Many biotas have consequently so adapted themselves to fire that, as with biotas frequented by floods and hurricanes, adaptation has become symbiosis. Such ecosystems do not merely tolerate fire, but often encourage it and even require it. In many environments fire is the most effective form of decomposition, the dominant selective force for determining the relative distribution of certain species, and the means for effective nutrient recycling and even the recycling of whole communities.

Evidence for the impact of human fires on vegetation is provided by the presence of charcoal in long sequences of lacustrine sediments. At Lynch's Crater, in the Atherton Tableland of Australia, for example, there is a sedimentary core that goes back 190 000 years. It is not until 38 000 years ago, however, that these sediments suddenly contain large quantities of charcoal, and thereafter there is evidence in the pollen record for an increase in the shrubby, small-leaved vegetation typical of regularly burnt habitats. These changes almost certainly coincide with, and result from, the first arrival of humans in Australia.

Thus, deliberately set fire, which doubtless often went out of control and covered huge areas, was the first major way in which quite small human groups modified major vegetation types and ecosystems.

Animal extinctions

As the size of human populations increased and technology developed, the greater would be the effect that hunters had on animal populations. The extent to which humans have been responsible for the extinction of many species of both birds and mammals has been argued about for a long time. It was for example, a problem that fascinated Alfred Russel Wallace (1914, pp. 248–9):

> It is sometimes thought that early man, with only the rudest weapons, would be powerless against large and often well-armed animals. But this, I think, is quite a mistake. No weapon is more effective for this purpose than the spear, of various kinds, when large numbers of hunters attack a single animal; and when made of tough wood, with the point hardened by fire and well sharpened, it is as effective as when metal heads are used. Bamboo, too, abundant in almost all

warm countries, forms a very deadly spear when cut obliquely at the point . . .

It is therefore certain, that, so long as man possessed weapons and the use of fire, his power of intelligent combination would have rendered him fully able to kill or capture any animal that has ever lived upon the earth . . . the numbers he would be able to destroy, especially of the young, would be an important factor in the extermination of many of the larger species.

Some workers, notably Martin (1967), believe that the human role in animal extinctions goes back to the Stone Age. Martin argues that the global pattern of extinctions of the large mammals follows the footsteps of human hunters. He suggests that Africa and parts of southern Asia were affected first, with substantial species losses at the end of the Acheulean, around 200 000 years ago. Europe and northern Asia were affected between 20 000 and 10 000 years ago, while Australia, and the Americas, the last major continental areas to which humans migrated, were stripped of large herbivores between 12 000 and 10 000 years ago. Extinctions continued into the Holocene after the retreat of the glaciers as large islands like Madagascar, Hawaii and New Zealand were colonised.

Such islands appear to have been especially prone to rapid and thorough ecological changes brought about by pre-European settlement. Large numbers of birds became extinct in Hawaii as Polynesian immigrants cleared lowland forests by fire. In New Zealand all moa species (perhaps 24 different species in all) and 21 other avian species were extinct before European settlement. Three factors contributed: the deforestation of one-third to one-half of the forested land surface by Polynesian fires; the direct hunting of the birds by Polynesian gastronomes; and the depredations of Polynesian dogs and rats. The moas were particularly susceptible to predation, being flightless, slow-moving, conservative in territoriality, complacent, reliant upon immobile males for egg incubation, and possessing little defensive ability beyond spasms of kicking.

The degree of extinction that took place in north-west Europe and the British Isles was also impressive (Stuart, 1982, p. 170):

Of the 13 species of large herbivores with body weights exceeding 200 kg, known from the Upper Pleistocene of Europe, 9 were extinct in Europe before the end of the Last Cold Stage. Of the 8 species with body weights over 600 kg (the rhinoceroses, *Bos*, *Bison*, *Hippopotamus* and the elephants), 7 were lost from the European fauna, and the sole survivor, *Bos primigenius* was exterminated, in its wild form, by man in historical times. Similarly, of the 4 largest Carnivora exceeding 50 kg body weight, known from the European Upper Pleistocene, 3 became extinct prior to the Flandrian; the exception being the brown bear *Ursus arctos*.

Arguments in favour of anthropogenic late-Pleistocene 'overkill' run as follows:

First, massive 'blitzkrieg' extinction in North America seems to coincide in time with the arrival of humans in sufficient numbers and with sufficient technological skill in making suitable artefacts (e.g. the famed Clovis bifacial blades) to be able to kill large numbers of animals.

Second, in Europe, sites such as La Solutré in France, where a late-Perigordian level (see the timescale in the Appendix to this chapter) is estimated to contain the remains of over 100 000 horses, attest to the efficiency of Upper Palaeolithic hunters. The Upper Palaeolithic in Europe (after *c*. 30 000 years before the present) saw a great expansion in technology, including pressure flaking, drilling, twisting, grinding, etc., and composite tools became more important. Thus whereas the Mousterian Neanderthals probably killed most of their prey by stabbing, with short spears with a large, broad stone point, the Upper Paleolithic *Homo sapiens sapiens* used spears with narrow sharp points with considerable penetrating power, reaching the parts other spears could not reach. Spear throwing and the bow-and-arrow were in use by 23 000 years before the present, permitting less dangerous and more stealthy hunting, even by individuals. Snares and traps were also developed, allowing a wider range of resources to be obtained.

Third, in addition to hunting and killing animals, humans may also have competed with them for particular food or water supplies. The latter might be crucial in the dry season of the seasonal tropics and arid lands, where game would be forced to concentrate near water-holes.

Fourth, many beasts unfamiliar with people are remarkably tame and naive in their presence, and it would have taken them time to flee or seek concealment at the sight or scent of human life.

Fifth, the supposed preferential extinction of the larger mammals could also lend support to the potency of human actions. Large body size means that big herbivores are required to feed almost continuously to sustain a large body mass. Moreover, as the size and generation time of a mammal increases so the relative rate at which it produces new tissue (whether as bodily growth or offspring) decreases. Hence a population of gazelle will turn over up to 70% of their biomass in a year, wildebeeste over 27%, but rhino and elephant only about 10%. The significance of this is that, since a population of large animals only turns over a small percentage of their biomass each year, the rates of slaughter that such a population can sustain in the face of even primitive hunters is very low indeed.

Sixth, the role of man is rendered likely by the fact that the fauna adjusted successfully, with relatively few casualties, to the very similar changes of climate and vegetation conditions during previous Cold Stage/Interglacial cycles, and there is no evidence for any unprecedented climatic event within the Last (Devensian) Cold Stage which could be linked to the extinction phenomenon.

Doubtless the great climatic changes at the transition from the Pleistocene to the Holocene, were also significant. There are considerable doubts surrounding such issues as to why small vertebrates also suffered large-scale late Pleistocene extinction in North America, why the North American bison survived in large numbers well into the nineteenth century, and why in

Australia humans and several species of large birds and mammals were living together for quite long periods before extinction occurred. Furthermore, as dates for human colonisation are pushed back, as, for example, by recent studies in South America, the correlations between dates of extinction and dates of human appearance in particular areas may become less strong.

It is probable that these extinctions had major ecological consequences. As Birks (1986, p. 49) has put it:

> The ecological effects of rapid extinction of over 75% of the New World's large herbivores . . . must have been profound, for example on seed dispersal, browsing, grazing, trampling and tree regeneration . . . Large grazers and browsers such as bison, mammoth and woolly rhinoceros may have been important in delaying or even inhibiting tree growth . . . giant deer in Ireland may have maintained the late-glacial treeless vegetation.

Agriculture and settlement

The third crucial development in environmental manipulation was the transition to agriculture and settled population – from savagery to barbarism. It is doubtful if the change was quite so sudden and revolutionary as was once thought, though Isaac (1970) has termed domestication 'the single most important intervention man made in his environment'. From as early as 18 000 years ago, some Nile Valley communities in Upper Egypt and Nubia may have been making relatively intense use of cereal foods, though the dates are still uncertain. Such use may have been accompanied by the care of wild grasses by such means as weed control, clearance of ground and, perhaps, occasional provision of water. Under these circumstances, the sowing of seeds might also have been employed to increase the density of growth in tended places and to extend the areas colonised by wild plants. With or without conscious selection for desired qualities, such practices would in the fullness of time have led to the development of crops which were morphologically distinct from their fully wild prototypes. Likewise, there may have been no clear-cut distinction between wild and domestic in the case of animals, for there are various stages in the process of the human exploitation of animals from random predation, through controlled predation, to herd following, to loose herding, and to close herding. Nonetheless, herding in the Zagros Mountains was being practised by 11 000–10 000 years ago, and crops such as emmer, einkorn and barley were being grown by much the same time.

Once established, agricultural technology developed rapidly. The earliest evidence of artificial irrigation is the mace-head of the Egyptian Scorpion King, which shows one of the last pre-dynastic kings ceremonially cutting an irrigation ditch around 5050 years ago.

Adverse environmental consequences of irrigation did not take long to manifest themselves (Fig. 1.3). Jacobsen and Adams have shown that salinity

Fig. 1.3. The introduction of irrigation around 5000 years ago probably caused substantial environmental changes in major river valleys in the Old World's arid areas. A major adverse consequence of rising ground water levels would be the initiation of water-logging and salinisation. This phenomenon occurs widely in the Indus plain of Pakistan today.

was a problem in Mesopotamian agriculture after about 4400 before the present, necessitating a substitution of barley for wheat and causing a severe decline in crop yields.

In the Old World the secondary applications of domesticated animals were explored. The plough was particularly important in this process – the first application of animal power to the mechanisation of agriculture. It was invented some 5000 years ago, and was used in Mesopotamia, Assyria and Egypt. The cart followed shortly thereafter and spread to Europe and India during the course of the third millennium. Use of the plough necessitated effective land clearance and also caused a new level of soil disturbance. Both tendencies would serve to increase the risk of soil erosion (Fig. 1.4).

Domestication, and these subsequent developments, reduced the space required for sustaining each individual by a factor of the order of 500 at least, and as a consequence we quickly see the establishment of the first human towns. The extent of Catal Huyuk in Anatolia (12.8 ha) and of Jericho (32 ha) suggest that urban status was reached by the sixth millennium BC or earlier. One should, nonetheless, probably not exaggerate the human population levels attained at this time. Flannery (1969) has suggested that the early dry farming systems may have had a carrying capacity of one to two persons per km^2, and early irrigated systems six or more persons per km^2 compared with

Fig. 1.4. Soil erosion is currently one of the most serious of all the environmental problems faced by the human race. The exposed roots of this tree from the Baringo area of northern Kenya, indicate the speed at which topsoil can be lost when the natural vegetation cover is removed by land clearance and overgrazing by domesticated animals.

0.1 persons per km^2 under late-palaeolithic hunter–gatherer systems. In other words, population could have increased more than 60-fold in the space of 6000 years. Wagstaff estimates that if around 5000 BC there were 35 500 villages of 100 inhabitants in the Middle Eastern region, each would require *c.* 42.4 ha of cereal-growing land, which amounts to 1.5 million ha (*c.* 2.3% of the area cultivated in 1971). Nonetheless, the process of urbanisation had developed, so that Nineveh may in due course have attained a population of 700 000, Carthage of 700 000 and Augustan Rome of around 1 000 000.

Post-glacial peoples in Britain

As the rigours of the Ice Age retreated, more and more portions of the British Isles became inhabited. Ireland was colonised by sea for the first time around 9000 years ago. Increasing evidence provided by pollen analysis indicates that the role of Mesolithic peoples was perhaps not inconsiderable, though Edwards and Ralston (1984) have rightly sounded a note of caution about its interpretation. In northern England in Mesolithic times (between about 10 000 and 5000 years ago) there was a seasonal pattern of transhumance with winter settlements in the lowlands and summer quarters in the uplands of the

Pennines or the North Yorkshire Moors. This increased the potential ecological impact, for the same bands were operating in two habitats differing in character and in spatial distribution. The pollen record shows that fire caused progressive deforestation in the Boreal. The dominance of *Corylus* (hazel) in the Boreal can itself be seen as the result of repeated burning. Being a plant which shoots freely from cut or burned stumps, hazel is not destroyed by fire (unless an intense ground fire develops) and so is able to sprout quickly and achieve the dominance that the pollen profiles show. This Boreal preponderance of hazel pollen is generally thought to be without parallel in the developing warm phase of any of the previous interglacials. The prevalence of open conditions at that time is reflected in the pollen record by the presence of light-demanding weeds, such as ribwort plantain (*Plantago lanceolata*) and sorrel (*Rumex*).

It is also possible that the expansion of alder (*Alnus glutinosa*) in Britain was partly due to human influence in the Mesolithic more than to increased wetness in the Atlantic period. The removal of the forest cover helped the spread of alder by reducing competition, as possibly did the burning of reed swamp. It may also have been assisted by deforestation and burning of stream catchment areas, which would lead to an increased run-off of waters. Furthermore, the felling of alder itself promotes vegetative sprouting and cloning which could result in its rapid spread in swamp forest areas.

The role of Neolithic peoples was even greater. A controversial issue here is the fall in elm (*Ulmus*) pollen. The so-called 'elm decline' appears approximately synchronously about 5000 years ago in all pollen diagrams from north-west Europe (though not from North America). Rackham (1980) believes that a simple deterioration of climate is inadequate to explain so sudden, universal and specific a change. It is possible that disease could have played a role, for Neolithic deposits on Hampstead Heath have revealed a prime carrier of the fungus which causes Dutch Elm Disease, the elm bark beetle *Scolytus scolytus*. An alternative explanation is that around 5000 years ago a new technique of keeping stalled domestic animals was introduced into Europe by Neolithic peoples, whose animals were fed by repeated gathering of heavy branches from nutritious elms. This, it was held, reduced the pollen production of the elms enormously. Experiments have shown that Neolithic peoples, equipped with polished stone axes, could cut down mature trees with some facility, and aided and abetted by fire, could clear a fair-sized patch of established forest within about a week. Such clearings were used for cereal cultivation. These 'Landnam' clearances, which were recognised by Iversen over 40 years ago (1941), involved the wholesale clearance of a patch of mixed forest by nomadic farmers practising shifting cultivation. The opening of the forest may have facilitated the entry or expansion of ash (*Fraxinus*), a light-demanding pioneer species. Its first appearance in most parts of the north and west of Britain coincides with the impact of Neolithic disturbance or of temporary clearance of primeval forest.

It is also possible that forest clearance and subsequent dereliction of clearings facilitated local colonisation and expansion of new immigrants such

Fig 1.5. In northern England it is possible that some of the bare expanses of limestone, exposed as 'pavements' in locations like Newbiggin Crags, may result from accelerated soil erosion brought about by Neolithic and later land clearance.

as beech (*Fagus sylvatica*), Norway Spruce (*Picea abies*) and possibly hornbeam (*Carpinus betulus*). Birks believes it likely that the rapid migration of *Fagus* across north-west Europe since 4000 before the present may have only been made possible by the creation of abundant, large clearings within oak (*Quercus*) or lime (*Tilia*) dominated forests on well-drained soils.

Domesticated sheep and cattle, together with wheat and barley fields, became a major feature of the British landscape in the fourth millennium BC, and 'within the astonishingly short space of 1000 years, had spread to all parts of the country as far north as Scotland' (Cunliffe, 1985, p. 52). The great density of earthwork structures such as causewayed camps, henge monuments, cursuses and long barrows implies a substantial and well-established population able to harness surplus manpower to undertake communal projects. Late-Neolithic henge monuments in southern England could have taken up to 1 million man-hours to construct, while Silbury Hill and Stonehenge might have taken 10–30 million man-hours respectively.

Forest clearance and large-scale use of wood caused major soil changes. One of these was accelerated erosion, which may have exposed some of our limestone pavements (Fig. 1.5). The evidence for this accelerated erosion is present in the Lake District where, in Neolithic times, there were changes in the stratigraphy and chemistry of lake sediments. At Barfield Tarn, in the West Cumbrian coastal plain, a change from an organic mud to a pink clay coincided with the 'elm decline'. The pink clay is interpreted as either a direct consequence of soil erosion resulting from oak–elm forest clearance and subsequent cultivation of boulder–clay soils around the tarn, or, at least, an acceleration of natural weathering by agricultural processes. Likewise, at Angle Tarn in the central mountains of the Lake District, chemical analyses of the sediments have provided evidence for the inwash of acid soil horizons. This indicates accelerated erosion of mineral soils following the clearance of forest on hill slopes.

Rather later, there is also increasing evidence to suggest that silty valley fills in Germany, France and Britain, many of them dating back to the Bronze Age and the Iron Age are the result of accelerated slope erosion produced by the activities of early farmers. The same applies to colluvial and alluvial sediments in Polynesia and Papua New Guinea. This too is a controversial matter, however, for there has been considerable debate in the literature about the relative importance of human activities and climatic changes in causing sediment erosion and deposition cycles of cut and fill in valley bottoms. Butzer (1974), for example, has propounded the importance of accelerated erosion in the Mediterranean lands, while Vita-Finzi (1969) has championed the role of climatic changes like the Little Ice Age.

A more controversial soil change is that involving acidification, the formation of strongly leached, nutrient poor podzol soils and peat bog development. Climatic changes and progressive leaching of Pleistocene drifts may have played a role, but the association in time and space of human activities with soil erosion is becoming increasingly clear. By replacing the natural woodland with fields and pastures, human societies set in train various related processes, especially on soils poor in minerals such as calcium which neutralise soil acidity. The destruction of deep-rooting trees curtailed the enrichment of the soil surface by minerals brought up from the deeper layers. Fire would release nutrients from vegetation in the form of readily soluble salts, some of which would inevitably be lost in drainage. The harvesting of crops and animal products would also lead to a loss in nutrients, while the invasion of bracken and heather would produce a more acidic humus type of soil than the original mixed deciduous tree cover. The development of podzols, by impeding the downward percolation of waters, may have accelerated the formation of peats. Perhaps of even greater importance is the fact that when a forest canopy is removed by deforestation the transpiration demand of the vegetation is reduced, less rainfall is intercepted, so that the supply of ground water is increased, aggravating any waterlogging or increasing run-off and consequent soil erosion (see Myers, this volume). In the southern Pennines the basal layers of all marginal peats contain widespread evidence of vegetation clearance by burning, either in the form of microscopic carbon particles (similar to present-day 'soot'), small charred plant fragments, or larger lumps of charcoal.

Thus, as Simmons (1981, p. 291) has said:

> . . . prehistoric peoples in Britain were influential changers of
> their ecology and their landscapes in their time, and even

afterwards where the later impress of man has not been so intensive. The major lineaments of the landscape of the uplands of today, in terms of their treelessness and much of their blanket bog, originated at the hands of prehistoric cultures, even if most government documents still refer to them as areas of 'natural' beauty. In the lowlands also, settlement patterns and field boundaries are in places still traceable to their prehistoric forebears. Even if environmental impact of prehistoric times has been dwarfed by more recent agricultural, urban and industrial processes, it is not to be overlooked in any serious account of the development of our landscape or the history of our ecology.

In spite of the increasing pace of world industrialisation and urbanisation, it is ploughing and pastoralism that are responsible for many of our most serious environmental problems and that are still causing some of the most widespread changes in the landscape. Thus soil erosion brought about by agriculture is, it can be argued, a more serious pollutant of the world's waters than is industry. Many of the habitat changes which so affect wild animals are brought about through agricultural expansion. Soil salinisation and desertification can be regarded as two of the most serious problems facing the human race (Fig. 1.6). Land-use changes, such as the conversion of

Fig. 1.6. Desertification, the spread of desert-like conditions to areas where they might not naturally occur, can be brought about by various types of human pressure. In this example from the Molopo Valley in Botswana, dunes have been rendered active as a result of overgrazing around a borehole.

[15]

forests to fields, are important factors causing anthropogenic changes in climate additional to the more celebrated burning of fossil fuels and emission of industrial aerosols into the atmosphere (see Bolin, this volume). The liberation of CO_2 into the atmosphere through agricultural expansion, changes in surface albedo (reflectivity) values and the production of dust, are all major ways in which agriculture may modify world climates (see Wells, this volume). These processes were set in train 10 000 years ago.

Above all it is pioneer agricultural colonisation of the rainforests by cattle keepers and simple cultivators, repeating in a sense the colonisation of the mid-European woodlands in the middle ages, or the overseas dominions in the nineteenth century, that presents one of the most daunting aspects of the human impact in the future, causing as it will rates of species extinction of an unparalleled magnitude. As Norman Myers (1979, p. 31) has expressed it: '. . . during the last quarter of this century we shall witness an extinction spasm accounting for 1 million species'. He has calculated that from AD 1600 to 1900, humans were accounting for the demise of one species every four years, that from 1900 onwards the rate increased to an average of around one per year, that at present the rate is about one per day, and that within a decade we would be losing one every hour.

Mineral exploitation

The next development in human cultural and technological life which was to increase human power for environmental modification was the mining of ores and the smelting of metals. This may have started around 5700 years ago, possibly in what is today north-west Iran, with the smelting of copper-oxide ores into metallic copper. The spread of metal working into other areas was rapid, and by 4500 years ago bronze products were in use from Britain in the west to northern China in the east. Metal working probably required enormous amounts of wood, and Sir William Flinders Petrie, in his investigation of the third millennium BC copper industry at Wadi Nash in Western Sinai, found a bed of wood ashes 30 m long, 15 m wide and 0.5 m deep. The smelting of iron ores may date back to as late as 1500 BC, and was probably discovered in western Asia. It was brought to north Africa in about the eighth century BC, and spread rapidly into Africa, reaching Natal in the first 300 years of the Christian era. The charcoal and wood requirements for iron-smelting are huge and deforestation took place on a substantial scale in Meroe and the ancient Kingdom of Ghana. Certain species of trees are especially desirable either because of their natural fluxing properties (e.g. *Terminalia* spp and *Combretum* spp) or because of their ability to generate high temperatures (e.g. *Acacia nilotica*, and *Tamarindus indicus*), and so may have been preferentially removed. Great erosional scars in Swaziland (Fig. 1.7), locally called *dongas*, are characterised by the presence of numerous furnaces and may have been caused by wholesale tree-clearance, which would have been further facilitated by the use of iron implements.

Fig. 1.7. In central Swaziland, great erosional scars, called *dongas*, often appear in conjunction with the remnants of ancient iron-working sites. The large amounts of charcoal required for iron smelting may have required the removal of protective vegetation over wide areas.

By the second century AD, Britain may have attained a population of 5 million and Wealden iron-smelting at Bardown probably led to the felling of 300–500 km^2 of Wealden forest by the middle of the third century.

Possibly the earliest widespread evidence for mining and mining landscapes in Britain comes from the flint mines of Grimes Graves in the Norfolk Breckland where there are the remains of some 600 mine-shafts (many up to 40 feet deep) sunk by Neolithic miners around 4700 years ago. Some 90 acres of heath are covered with hummocks and funnel-shaped hollows.

Plainly, a whole new suite of environmental problems has emerged as a result of the exploitation of minerals in recent centuries: the burning of coal and oil contributes to the build-up of CO_2 levels in the atmosphere and the acidification of precipitation; the widespread development of chemicals from hydrocarbons, including synthetic fertilisers, makes a major addition to the potential for air and water pollution; heavy metals and radioactive elements pose problems for human life; smelters, such as that at Sudbury in Ontario, pour noxious gases into the atmosphere.

However, as I have tried to argue in this chapter, the importance of human agency in environmental change has been long continued. Thus although much of the concern expressed about the undesirable effects humans have tends to focus on the role played by sophisticated industrial societies, this should not blind us to the fact that many highly significant environmental changes were and are being achieved by non-industrial societies. Even

Table 1.2. *Requirements of living space (area per head, km²)*

Location	Economy	Area
Australia	Hunting	30–40
Tasmania		17
New Zealand (pre-Moa extinction)		10–15
New Zealand (post-Moa extinction)		30
NW Canada		140
US Prairies		20–25
E Africa (Masai)	Pastoralism	0.6–1.3
Nigeria (Fulani)		1.1
E Africa (Turkana)		0.8
E Africa (Somalis)		0.5
Philippines	Shifting agriculture	0.03
Sarawak		0.5
Zambia		0.1
Ivory Coast		0.1
Guatemala		0.01
Sumatra		0.06–0.025

Source: From information in Clark & Haswell (1970).

though the population densities of human hunter and gatherer predators are necessarily low (Table 1.2), they have achieved substantial modifications of some major vegetation types through the use of fire. Furthermore, as the technological ability of such groups evolved, notably in the upper Palaeolithic, a major spasm of animal extinctions was achieved, wiping out many of the world's largest and most ferocious beasts. When the domestication of plants and animals occurred, adoption of such relatively extensive land use practices as pastoralism and shifting agriculture caused a great potential for a reduction in the requirements for living space compared to hunting and gathering activities. When irrigation and ploughing were introduced, another threshold of population densities was achieved and there is good early evidence for accelerated erosion and salinisation in many parts of the world caused by prehistoric societies. Such impacts remain among the most serious and intractable that we face today. As W.C. Lowdermilk wrote in 1935, 'the lands of the earth are occupied: frontiers of new lands have disappeared. The only new frontier that appears is underfoot, in the maintenance of productivity of lands now occupied.'

Appendix

Timescale of events referred to in the text

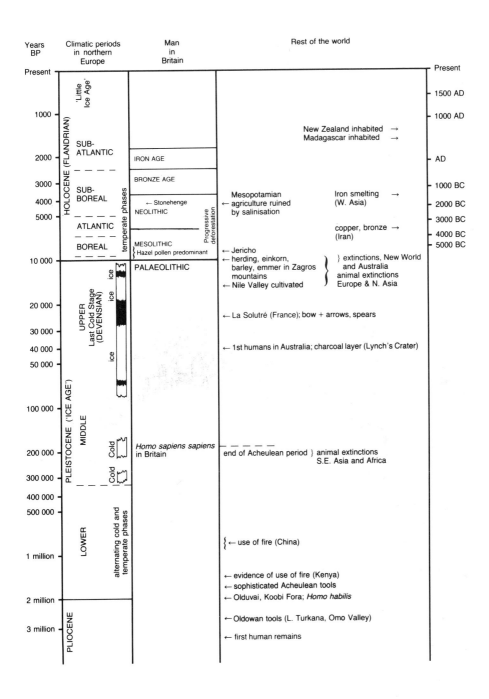

Years BP	Climatic periods in northern Europe	Man in Britain	Rest of the world	
Present				Present
	'Little Ice Age'			1500 AD
1000	HOLOCENE (FLANDRIAN) SUB-ATLANTIC			1000 AD
			New Zealand inhabited → Madagascar inhabited →	
2000		IRON AGE		AD
3000		BRONZE AGE		1000 BC
4000	SUB-BOREAL	← Stonehenge NEOLITHIC	Mesopotamian ← agriculture ruined by salinisation	Iron smelting → (W. Asia)	2000 BC
5000	ATLANTIC			copper, bronze → (Iran)	3000 BC
	BOREAL	MESOLITHIC Hazel pollen predominant		4000 BC / 5000 BC
10 000		PALAEOLITHIC	← Jericho ← herding, einkorn, barley, emmer in Zagros mountains ← Nile Valley cultivated	} extinctions, New World and Australia animal extinctions Europe & N. Asia
20 000	UPPER Last Cold Stage (DEVENSIAN)		← La Solutré (France); bow + arrows, spears	
30 000				
40 000			← 1st humans in Australia; charcoal layer (Lynch's Crater)	
50 000				
100 000	PLEISTOCENE ('ICE AGE') MIDDLE			
200 000		*Homo sapiens sapiens* in Britain	end of Acheulean period } animal extinctions S.E. Asia and Africa	
300 000				
400 000				
500 000	LOWER			
1 million			{ ← use of fire (China)	
			← evidence of use of fire (Kenya) ← sophisticated Acheulean tools	
2 million			← Olduvai, Koobi Fora; *Homo habilis* ← Oldowan tools (L. Turkana, Omo Valley)	
3 million	PLIOCENE		← first human remains	

Man in Britain column includes: "temperate phases" and "Progressive deforestation" (NEOLITHIC/ATLANTIC section); "ice" markings in Devensian section; "alternating cold and temperate phases" and "Cold" markings in Pleistocene section.

Further reading

Bell, M. (1982). The effects of land-use and climate on valley sedimentation. In A. Harding (ed.) *Climate Change in Later Prehistory*. Edinburgh: The University Press, pp. 127–42.

Birks, H.J.B. (1986). Late-Quaternary biotic changes in terrestrial and lacustrine environments, with particular reference to north-west Europe. In B.E. Berglund (ed.) *Handbook of Holocene Palaeoecology and Palaeohydrology*. Chichester: Wiley, pp. 3–65.

Butzer, K.W. (1974). Accelerated soil erosion: a problem in man-land relationships. In I.R. Manners & M.W. Mikesell (eds.) *Perspectives on Environments*. Washington DC: Association of American Geographers.

Clark, C. & Haswell, M. (1970). *The Economics of Subsistence Agriculture* (4th edn). London: Macmillan, 267 pp.

Clarke, J.D. (1985). Leaving no stone unturned: archaeological advances and behavioural adaptation. In P.V. Tobias (ed.) *Hominid Evolution: Past, Present and Future*. New York: A. Liss, pp. 65–88.

Clarke, J.D. & Harris, J.W.K. (1985). Fire and its roles in early hominid lifeways. *African Archaeological Review*, **3**, 3–27.

Cunliffe, B.W. (1985). Man and landscape in Britain 6000 BC–AD 400. In S.R.J. Woodell (ed.) *The English Landscape Past, Present and Future*. Oxford: Oxford University Press, pp. 48–67.

Dennell, R. (1985). *European Economic Prehistory. A new approach*. London: Academic Press, 217 pp.

Edwards, K.J. & Ralston, I. (1984). Postglacial hunter-gatherers and vegetational history in Scotland. *Proceedings of the Society of Antiquaries, Scotland*, **114**, 15–34.

Flannery, K.V. (1969). Origins and ecological effects of early domestication in Iran and the Near East. In P.J. Ucko & G.W. Dimbleby (eds.) *The Domestication and Exploitation of Plants and Animals*. London: Duckworth, pp. 73–100.

Gillieson, D., Oldfield, F. & Krawiecki, A. (1986). Records of prehistoric soil erosion from Rock-Shelter sites in Papua New Guinea. *Mountain Research and Development*, **6**, 315–24.

Goudie, A.S. (1986). *The Human Impact on the Environment* (2nd edn). Oxford: Basil Blackwell.

Gowlett, J.A.J., Harris, H.W.K., Walton, D. & Wood, B.A. (1981). Early archaeological sites, hominid remains and traces of fire from Chesowanja, Kenya. *Nature*, **294**, 125–9.

Haaland, R. (1985). Iron production, its socio-cultural context and ecological implications. In R. Haaland & P. Shinnie (eds.) *African Iron Working – Ancient and Traditional*. Oslo: Norwegian University Press, 209 pp.

Isaac, E. (1970). *Geography of Domestication* Englewood Cliffs: Prentice Hall.

Iversen, Z. (1941). Landnam i Danmarks Stenalder. *Danmarks geologiske Undersogelse*, series 11, **66**, 68 pp.

Jacobsen, T. & Adams, R.H. (1958). Salt and silt in ancient Mesopotamian agriculture. *Science*, **123**, 1251–8.

Kershaw, A.P. (1986). Climatic change and aboriginal burning in north-east Australia during the last two glacial/interglacial cycles. *Nature*, **322**, 47–9.

Kirch, P.V. (1983). Man's role in modifying tropical and subtropical Polynesian ecosystems. *Archaeology in Oceania,* **18,** 26–31.

Lowdermilk, W.C. (1935). Man-made deserts. *Pacific Affairs,* **8,** 409–19.

Marsh, G.P. (1864). *Man and nature.* New York: Scribner.

Marshall, L.G. (1984). Who killed Cock Robin? An investigation of the extinction controversy. In P.S. Martin & R.G. Klein (eds.) *Quaternary extinctions.* Tucson: University of Arizona Press.

Martin, P.S. (1967). Prehistoric overkill. In P.S. Martin & H.E. Wright (eds.) *Pleistocene extinctions.* New Haven: Yale University Press, pp. 75–120.

Myers, N. (1979). *The Sinking Ark: a New Look at the Problem of Disappearing Species.* Oxford: Pergamon Press.

Pyne, S.J. (1982). *Fire in America – A Cultural History of Wildland and Rural Fire.* Princeton: Princeton University Press.

Rackham, O. (1980). *Ancient Woodland.* London: Arnold.

Sauer, C.O. (1969). *Seeds, Spades, Hearths and Herds.* Cambridge, Mass.: MIT Press.

Sherratt, A. (1981). Plough and pastoralism: aspects of the secondary products revolution. In I. Hodder, G. Isaac & N. Hammond (eds.), *Pattern of the Past.* Cambridge University Press, pp. 261–305.

Simmons, I.G. (1981). Culture and environment. In I.G. Simmons & M.J. Tooley (eds.), *The Environment in British Prehistory.* London: Duckworth, pp. 282–91.

Stuart, A.J. (1982). *Pleistocene Vertebrates in the British Isles.* London: Longman, 212 pp.

Vita-Finzi, C. (1969). *The Mediterranean Valleys.* Cambridge University Press.

Wagstaff, J.M. (1985). *The Evolution of Middle Eastern Landscapes. An Outline to AD 1840.* London: Croom Helm, 304 pp.

Wallace, A.R. (1914). *The World of Life. A Manifestation of Creative Power, Directive Mind and Ultimate Purpose.* London: Chapman and Hall, 408 pp.

[2]

The future of forests

Norman Myers

Introduction

Forests are arguably the most important vegetation zone on the face of the earth today. They play a far greater role in the well-being of the planetary ecosystem than they are often given credit for, and we may soon find that forests will effectively be called upon to make a still more critical contribution to planetary stability. Yet on every side, from the equator to the arctic, forests are being depleted or will shortly be depleted through human agency at a rate that could well reduce many of them to impoverished remnants by the end of the next century. Indeed, forests, which have been the predominant form of vegetation on our planet for hundreds of millions of years, may soon become a minority presence – whereupon we shall discover (by default) the full measure of their part in underpinning the ecological welfare of our biosphere.

What is the nature and scope of the role of forests in planetary workings? By 'forests' I mean tree-dominated communities with substantial canopies, Fig. 2.1 (by contrast with woodlands which feature much sparser tree cover) that are the repository of a greater abundance and diversity of terrestrial life forms than the rest of the earth put together. Tropical forests are specially rich in species and in the evolutionary capacity to generate new species. As tropical forests are cleared wholesale, there will be an impoverishing impact on the very course of evolution itself.

Moreover, forests help to regulate the hydrodynamics of great watersheds and river basins such as those of the Ganges and the Amazon. When forests are cut, particularly on sloping land, water runs off unimpeded, taking large quantities of soil and nutrients with it, thus depleting the agricultural potential of the deforested area and causing increased flooding and silt accumulation downstream. Forests also help to regulate climate, at least at local level, through their moisture-cycling processes and their reduction of the reflectivity of the earth's surface (albedo) (see Bolin, this volume). Forests harbour much more carbon in the form of plant tissue, detritus and soil than the rest of the planet put together (Table 2.1), so they play a critical role in the global carbon budget. As forests are eliminated by burning, they serve as an anthropogenic source of carbon dioxide, thus contributing to the 'greenhouse

Fig. 2.1. Tropical rainforest in Panama (United Nations).

Table 2.1. *Forests of the world, areas and carbon stocks*

Forest type	Area (millions km^2)	Carbon stocks in plants (gigatons, or billion metric tons, of carbon)	Carbon stocks in detritus and soil (gigatons, or billion metric tons, of carbon)
Tropical forests (both evergreen and seasonal, and in primary form)	8.6	202	288
Temperate-zone forests	8.2	65	161
Boreal forests	11.7	127	247
Woodland and shrublands	12.8	57	59
Rest of Earth's land surface	103.3	109	720
Totals	144.6	560	1475

Data from B. Bolin *et al.* (1986).

effect', a phenomenon that may transform our planet to a profound degree within just another few decades (see Bolin, this volume). In various ways, then, forests exert a great gyroscopic effect in maintaining the dynamic stability (homeostasis) of many biospheric functions, including many functions that keep the planet capable of sustaining life as we know it.

In this chapter, we shall look at ways in which forests, from tropical forests to temperate-zone and boreal (frost hardy) forests, are being depleted. (The present extent of these three types of forest is given in Table 2.1 and Fig. 2.2) Which human communities are causing the damage? How are they doing it? How fast and with what specific consequences?

Since tropical forests are far richer in species and life-forms than other forests, and since they are undergoing much more rapid depletion, we shall start by considering this major vegetation type.

Tropical forests

Rates and trends of deforestation

Tropical forests in primary (undisturbed) form cover some 8.6 million km^2, all that remains of 15–16 million km^2 that may once have existed, according to bioclimatic data. There is general agreement that between 76 000 and 100 000 km^2 of these forests are eliminated outright each year; and that at least

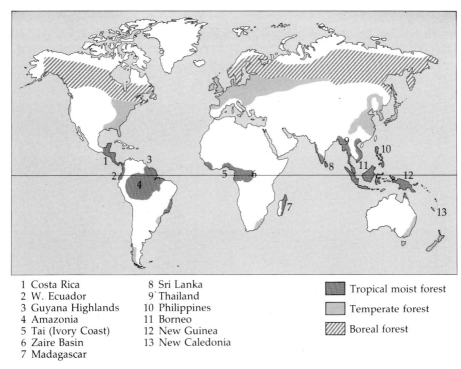

1 Costa Rica	8 Sri Lanka	
2 W. Ecuador	9 Thailand	
3 Guyana Highlands	10 Philippines	
4 Amazonia	11 Borneo	
5 Tai (Ivory Coast)	12 New Guinea	
6 Zaire Basin	13 New Caledonia	
7 Madagascar		

Tropical moist forest

Temperate forest

Boreal forest

Fig. 2.2. World distribution of various types of forest.

a further 100 000 km^2 are grossly disrupted each year (for details of references, see the end of this chapter). Moreover, these figures for depletion rates derive from a data base of the late 1970s, and the rates have increased somewhat since then. All this means, roughly speaking, that 1% of the tropical forest biome is being deforested each year, and rather more than another 1% is being significantly degraded. By the end of this century, or shortly thereafter, there could be little left of tropical forest of undisturbed primary status, with full biotic diversity and ecological complexity, outside of two large remnant blocs: one in the Zaire Basin and the other in the western half of Brazilian Amazonia, plus some outlier areas in the Guyana highlands and in New Guinea. While these relict areas of tropical forest may well endure for several decades longer, there is little likelihood of them lasting beyond the middle of the next century, if only because of the sheer expansion in numbers of small-scale cultivators.

The developed world plays a major role in tropical deforestation. Through our consumer practices, our own hand is on the chainsaw. This is principally apparent in the international trade in tropical timber and the overexploitation of forests by commercial loggers (Fig. 2.3). The excessive harvest is stimulated in major measure by developed-world demand for tropical hardwoods. Of course, developing countries themselves consume much of the hardwood, but developed-world consumption has increased 15 times since 1950,

Fig. 2.3. Commercial logging in Sabah, Malaysia, 1983 – an example the destruction of tropical forest (Seaphot Ltd., photo by Richard Matthews).

whereas producer-region consumption has increased only three times, until each is now responsible for about half the total.

The most notable instance of developed-world consumption of tropical hardwoods is Japan, which takes three-quarters of timber exports from the region undergoing the most rapid deforestation, southeast Asia (Fig. 2.3). Yet Japan could supply its entire hardwood needs from its own domestic forests, since two-thirds of the country is covered with good-quality forests. But annual removals from Japan's forests have been reduced by half during the last 20 years, until they now represent considerably less than annual renewal by growth. Indeed, each year Japan continues to bring yet more of its hardwood forests under protected status, on the grounds that it can still take advantage of 'cheap' supplies of hardwoods from the humid tropics. However laudable Japan's domestic forestry policies may be within its national context, they are far less commendable within a global perspective, insofar as they foster misuse and overuse of timber stocks in tropical forests.

Similar considerations, though not nearly on the same scale, apply to Britain, which imports 90% of its timber needs, a good part of them from the tropics. All in all, Britain accounts for about one-eighth of developed-world imports of hardwoods from the tropics. This throws light on the current controversy about reforestation practices in Britain, which are resisted by conservationists who prefer open windswept moorlands to 'plantation plastered' landscapes.

A further role of developed-world consumerism is seen in the 'hamburger connection': demand for cheap beef from North American consumers fosters the spread of cattle ranches in Central America, at the cost of forest cover. A similar link can be recognised in western Europe, in the form of the 'cassava connection': cassava is extremely rich in calories, hence its demand as a foodstuff for livestock in the European Economic Community has increased, thus contributing to the 'beef mountain'. The Community has been absorbing about 80% of internationally traded cassava, most of the supplies being grown in Thailand (where cassava is not a traditional food crop). Since cassava is available at comparatively cheap prices in Thailand, it is a commercially attractive supplement for Europe's livestock feed. In 1973 the Community imported 1.5 million metric tons of cassava from Thailand, an amount that rapidly rose to more than 8 million tons by 1982. Today, Thailand enjoys a guaranteed quota from the Community of 5.5 million tons a year, which enables it to account for about 90% of cassava exports worldwide.

The principal cassava growers in Thailand include a fast-increasing number of small-scale farmers, especially in the eastern and north-eastern parts of the country where they establish their crops at the expense of natural forest. It is these parts of Thailand that are suffering the most rapid decline in their forest cover, and where environmental repercussions, such as soil impoverishment and erosion, plus disruption of watershed systems, are most severe.

Extinction of species

Tropical forests cover only 6% of earth's land surface, yet they harbour at least 50% of all earth's species. Indeed, some recent research suggests that just the canopy of tropical forests may contain 30 million (conceivably 50 million) insect species alone (see May, this volume). Of course, these are no more than bald statistics, and they do not convey an idea of the extreme biotic diversity of tropical forests. In ten 1-ha plots in Borneo, 700 species of trees have been documented, or as many as in the whole of North America. A single locality of Costa Rica, the La Selva Forest Reserve, covering a mere 13.7 km^2, features more than 1800 vascular plant species, 394 breeding birds, 104 mammals, 76 reptiles, 46 amphibians, 42 fish and 143 butterflies. This compares with the whole of Great Britain, 233 000 km^2 with around 1400 native plant species, about 240 breeding birds, 47 mammals, 6 reptiles, 6 amphibians, 43 fish, and 64 butterflies. On a single leguminous tree in the Tambopata Reserve in Amazonian Peru have been found 43 species of ants belonging to 26 genera, or roughly as many as in the entire ant fauna of the British Isles.

It is tropical forest depletion that is far and away the main cause of the species extinction spasm that is overtaking the biosphere. To gain an insight into the scope and scale of extinctions in tropical forests, let us briefly consider three particular areas: western Ecuador, Atlantic-coast Brazil and Madagascar. Each of these areas features, or rather did feature until recently, exceptional concentrations of species in its forests, with a large proportion of endemic species, *i.e.* unique to that area. As Dr Alwyn Gentry has documented, western Ecuador is reputed to have once contained between 8000 and 10 000 plant species, of which somewhere between 40 and 60% are endemic. If we suppose, as we reasonably can by drawing on detailed inventories in sample plots, that there are between 10 and 30 animal species for every one plant species, the species complement in western Ecuador must have amounted to about 200 000 in all. Since 1960, at least 95% of the forest cover has been destroyed, to make way for banana growing, oil exploitation and human settlements of various sorts. Using biogeographical techniques to relate the number of species likely to be found in a habitat to its area, we can realistically reckon that when a habitat has lost 90% of its extent, it has lost half its species. Precisely how many species have actually been eliminated, or are on the point of extinction, in western Ecuador is impossible to say. But ultimate accuracy is surely irrelevant, insofar as the number must total tens of thousands at least, conceivably 50 000 – all accounted for in the space of just 25 years.

Very similar baseline figures for species totals and endemism levels, and a similar record of forest depletion (albeit for different reasons, and over a longer time period), apply to the Atlantic-coastal forest of Brazil, where the original 1.25 million km^2 of forest cover have been reduced to less than 50 000 km^2. Parallel data also apply to Madagascar, except that the proportion of endemic species is a good deal higher. So in these three tropical forest areas alone, with their roughly 600 000 species, we must be witnessing a wave of extinctions.

As for the future, the outlook seems all the more adverse, though its detailed dimensions are even less clear than those of the present. In addition to the three critical areas listed, we can identify several other sectors of the tropical forest biome that feature exceptional concentrations of species with unusually high levels of endemism, and that face severe threat of depletion. They include the Choco forest of Colombia; the Napo centre of diversity in Peruvian Amazonia, plus several other centres that lie around the fringes of the Amazon basin and hence are unusually threatened by settlement programmes and various other forms of development; the Tai Forest of the Ivory Coast; the montane forests of East Africa; the relict wet forest of Sri Lanka; the monsoon forests of the Himalayan foothills; north-western Borneo; certain sectors of the Philippines; and several islands of the South Pacific. (New Caledonia, for instance, with 16 325 km^2, or almost the size of Wales, contains 3000 plant species, 80% of them endemic.)

These sectors of the tropical forest biome amount to roughly 1 000 000 km^2 (four times the size of Gt Britain), or slightly more than one-tenth of remaining undisturbed forests. So far as can best be judged from their documented numbers of plant species, and by making substantiated assumptions about the numbers of associated animal species, these 20 areas must surely harbour one million species (see May, this volume). Many of the areas contain species which occur nowhere else and which will disappear completely if their habitat is destroyed. If present land-use patterns and exploitation trends persist (they show every sign of accelerating), there will be little of these forest tracts left, except in the form of degraded remnants, by the end of this century or shortly thereafter. Thus forest depletion in these areas alone could well eliminate large numbers of species, surely hundreds of thousands and possibly many more, within the next 25 years.

What is the prognosis for the longer-term future? Eventually we could lose at least one-quarter, possibly one-third and conceivably a still larger share of all earth's species, the great bulk of the extinctions taking place in tropical forests. Let us take a quick look at the case of Amazonia, and the calculations of Professor Daniel Simberloff (see references). If deforestation in Amazonia continues at present rates until the year 2000, but then comes to a complete halt, we would anticipate (from biogeographical predictions) a loss of about 15% of plant species, and a similar share of animal species. Were the forest cover to be ultimately further reduced to those areas now set aside as parks and reserves, we would anticipate that 66% of plant species would eventually disappear, together with almost 69% of bird species, and similar proportions of all other major categories of species.

Indeed it is not going too far to say that the mass-extinction episode underway in tropical forests, with the numbers of species involved and the telescoped time-scale of the phenomenon, may result in the greatest single setback to life's abundance and diversity since the first emergence of life almost four billion years ago.

However deplorable may be this wholesale destruction of vast throngs of species with which we share this planet, yet still more significant could be the impoverishing impact on the future course of evolution. To cite the graphic phrasing of Drs Michael E. Soule and Bruce A. Wilcox of California: 'Death is one thing: an end to birth is something else.' From the little we can discern from the geologic record, the 'bounce-back' recovery time may amount to millions of years. After the demise of the dinosaurs 65 million years ago, somewhere between 50 000 and 100 000 years elapsed before there started to emerge a set of biotas as diversified and specialised as those which had been there before. A further 5–10 million years went by before there were bats in the skies and whales in the seas. So the mass-extinction phenomenon we are imposing on the biosphere could generate an impoverishing impact on the future course of evolution itself.

By contrast with other mass-extinction episodes of the geologic past, the critical factor this time lies with the likely loss of key environments. We appear set to lose most if not virtually all of the tropical forest biome; and this ecological zone has served in the past as a pre-eminent 'powerhouse' of evolution, generating more species than other environments. It has long been thought that virtually every major group of vertebrates and many other large categories of animals have originated in regions with warm, equable climates, notably the Old World tropics, and especially their forests. It has likewise been supposed that the rate of evolutionary diversification, whether through proliferation of species or through emergence of major new adaptations, has been greatest in the tropics, and particularly in tropical forests. In addition, tropical species, especially tropical forest species, appear to persist for only brief periods of geological time, which implies a high rate of evolution and extinction.

Of course tropical forests have been severely depleted in the past. During drier phases of the late Pleistocene they were repeatedly reduced to only a small fraction, occasionally as little as one-tenth, of their former expanse. But the remnant forest 'refugia' usually contained sufficient stocks of surviving species to recolonise suitable territories when moist conditions returned. Within the foreseeable future, by contrast, it seems all too possible that most tropical forests will be reduced to much less than one-tenth of their former expanse, and their pockets of surviving species will be so much less stocked with potential colonisers.

Furthermore, the species depletion will surely apply across most, if not all, major categories of species. This is almost axiomatic if extensive environ-ments, such as those of tropical forests, are eliminated wholesale. So the result will contrast sharply with the disappearance of the dinosaurs at the end of the Cretaceous, when not only placental mammals survived (leading to the adaptive radiation of mammals, eventually including man), but also birds, amphibians, crocodiles and many other non-dinosaurian reptiles. The present extinction episode looks likely to eliminate a sizeable share of terrestrial plant species, at least one-fifth within the next half-century and a good many more within the following half-century. During most mass-extinction episodes of the prehistoric past, by contrast, terrestrial plants have survived with relatively few losses, providing a continued source of food and shelter for the replacement animal species generated by subsequent evolutionary processes. If this resource is markedly depleted within the foreseeable future, the restorative capacities of evolution will be diminished all the more.

The impending upheaval could rank as one of the greatest biological revolutions of all time. In scale and significance, it could equal the development of aerobic respiration, the emergence of flowering plants, and the evolution of limbed animals. But whereas these three examples rank as advances, the wholesale destruction of so many species would, of course, rank as a distinct setback. We are the first species ever to be able to look out upon the biosphere, and to decide whether we would remake part of it – to consciously determine the future course of evolution.

Elimination of genetic resources

There is a further aspect to mass extinction of species, and that is the broadscale elimination of gene reservoirs. The genetic variability of species makes many contributions to modern agriculture, medicine and industry. Moreover, they make these contributions after scientists have conducted intensive investigations into only one plant species in 100, and a far smaller proportion of animal species. If scientists were to engage in comprehensive and systematic screening of species for their economic applications, we could surely look forward to entire cornucopias of new and improved foods, whole pharmacopoeias of new drugs and medicines, and unusually diversified stocks of raw materials for innovative industry. However utilitarian this argument in support of species conservation may appear in the eyes of some observers, it is an economic rationale that appeals in some quarters more than do aesthetic or ethical arguments. This applies particularly to the developing world, where the biggest concentrations of species exist, where extinction threats are greatest, and where conservation capacities are in shortest supply.

The diverse contributions of genetic resources to our material welfare have been widely documented in the past few years. The case of a species of wild maize, recently discovered in a montane forest of south-central Mexico, is instructive. This plant is the most primitive known relative of modern maize and, at the time of its discovery, it was surviving in only three tiny patches, covering a mere four hectares. Its habitat was threatened with imminent destruction by squatter cultivators and commercial loggers. The wild species turns out to be a perennial, unlike all other forms of maize, which are annuals. Now that it has been cross-bred with established commercial varieties, it opens up the prospect that maize growers (and maize consumers) could be spared the seasonal expense of ploughing and sowing, since the plant would spring up again of its own accord, like grass or daffodils.

Even more important, the wild maize offers resistance to at least four of eight major viruses and mycoplasmas that have hitherto baffled corn breeders. These four diseases cause at least a 1% loss to the world's maize harvest each year, worth more than $500 million. Still more to the point, the wild species, discovered at elevations between 2500 and 3300 metres, is adapted to habitats that are cooler and damper than established maize-growing lands. This offers scope to expand the cultivation range of maize by as much as one-tenth. All in all, the genetic benefits supplied by this wild plant, recently surviving in the form of no more than a last few thousand stalks, could possibly be worth a total of several billion dollars per year.

Tropical forest species likewise contribute to our health needs. On average, one out of four medicines available on prescription – whether antibiotic, tranquilliser, diuretic, laxative, or contraceptive pill – owes its origin to raw materials from wild plants of tropical forests. If non-prescription items are included, the commercial value of all products derived from tropical forest plants now amounts worldwide to some $15 billion a year.

The best-known illustration is that of the rosy periwinkle, a plant originally from Madagascar's forests (where, as we have noted, perhaps thousands of plant species have recently become extinct or are on the point of extinction). Alkaloids produced from this plant have been used widely for chemotherapy in certain cancers. According to the National Cancer Institute in the United States, there could well be another five plants in Amazonia alone with capacity to generate superstar drugs against cancer.

As for animals, a tropical forest species, the African green monkey (*Cercopithecus aethiops*), appears to be frequently infected by a virus very similar to the one that causes AIDS in humans, yet remains largely unaffected. According to some recent investigations, the monkey may provide a uniquely useful model for the study of AIDS, conceivably leading to the eventual development of a vaccine.

Many industrial benefits are also derived from tropical forest species, in the form of gums and exudates, essential oils and ethereal oils, resins and oleoresins, dyes and tannins, vegetable fats and waxes, insecticides and multitudes of other compounds. Many wild plants bear oil-rich seeds with potential for manufacture of fibres, detergents, starch and general edibles – even for an improved form of golf ball. Especially promising are certain *Fevillea* species, these being vines of western Amazonia that feature a higher oil content than is the case for any other dicotyledonous plant. If naturally occurring lianas in a tropical forest were to be cut and replaced by *Fevillea* plants, there could be a per-hectare oil yield comparable to yields obtained in the most productive oil-palm plantations, and they could be obtained without felling a single forest tree.

Ironically, the greatest immediate losers through species extinctions and depletion of gene reservoirs in tropical forests are the developed nations. These nations possess much more technological capacity to exploit genetic resources, especially through the emergent techniques of genetic engineering, than do developing nations. Again, we can perceive a linkage between the fortunes of developed nations far outside the tropics, and those of tropical-forest nations themselves.

Climatic dislocations

Finally, let us look at a crucial and hotly disputed aspect of tropical deforestation. There is growing evidence that at regional levels at least, a decline in forest cover can be closely associated with a decline in rainfall regimes. We know all too little about the interactions, if any, at work. Yet if any causative relationship exists, this could generate some unusually widespread and adverse impacts if, for instance, it were to lead to reduced stocks of moisture for agriculture. A seemingly negligible decline in rainfall, amounting to a few percent, can sometimes have major effects on crop yields.

'Does deforestation have the capacity to change climate?' This question is increasingly asked, yet climatologists are far from able to supply a definitive answer. The best they can offer is a series of insights into the linkages involved. In certain localities and under certain circumstances, there is higher rainfall in forest areas, but this does not mean that forests increase rainfall. One of the wettest places anywhere, Cherrapunji, with an annual average of 11.5 metres of rainfall, no longer possesses much forest cover at all, yet rainfall does not seem to have changed in amount; rather it runs off more rapidly, for reasons given below, leading to chronic shortages of drinking water. Similarly, evidence that forests do not appear to induce rain does not mean that forests *cannot* induce rain.

Certain observers insist there is little or no effect of forest cover on rainfall regimes. But a good number of others believe there is, or could well be, a connection, even a substantial connection. In essence, the putative inter-action amounts to this. Forests generally exchange moisture and energy with the atmosphere more intensively than do any other types of land-surface cover. The main mechanism lies with evapotranspiration, the loss of water drawn up from soil and evaporated from leaves, which in turn depends on three requirements: moisture in the soil; vegetation that transfers moisture from the soil to the atmosphere; and energy that converts the moisture into water vapour. Most of the energy comes from radiational heating of the surface, and thus depends on surface albedo (or 'shininess' of the land surface). In turn, the albedo depends on the vegetation, which in turn again depends on the soil moisture. As vegetation is eliminated, so there is less scope for evapotranspiration, and less moisture is dispatched into the atmosphere for recycling as rain. Human modification of vegetation cover could well bear significantly on climate, if it is of large enough scale and scope (with the exact response being subject to a good deal of variation, depending upon local conditions).

To see how all this works out in practice, consider the case of Amazonia. Studies using water isotopes reveal that between one-half and four-fifths of the region's moisture remains within the ecosystem. That is, it is constantly transpired by plants into the atmosphere, where it gathers in storm clouds before being precipitated back onto the forest. According to the pioneering research of Professor Eneas Salati and his colleagues, the forest thus represents a significant source of its own moisture. Were Amazonia to be widely deforested, there would be a sizeable decrease in the amount of moisture being evapotranspired into the atmosphere, eventually leading to a decline in rainfall of one-quarter or even more. A rainfall decline of this order would entrain some marked and irreversible ecological changes in many areas. Even more important, it could trigger a self-reinforcing process of growing desiccation of the forest cover, with declining moisture stocks followed by yet more desiccation, and so forth. Eventually, the repercussions could extend outside Amazonia, even to southern Brazil with its major agricultural lands.

Next, let us consider the Ivory Coast, where somewhat different mechan-isms may well be at work. There has been progressive replacement of forest with croplands during the past 30 years. There has also been a steady decline in evapotranspiration rates, from 60 to 35%, i.e. from a level typical of forest to a level typical of savannah, while run-off has increased eight-fold. Although climate in the Ivory Coast is more affected by nearby oceanic influences than is the case in Amazonia, nonetheless there seems to have been an association between deforestation and drying-out of formerly forested lands during a period of 30 years. Areas that were formerly excellent for cocoa growing are being abandoned because of less rainfall, less humidity, and longer and harsher dry seasons, plus lowering of water tables and drying up of wells.

A similar association is apparent in the Panama Canal area, where steady deforestation since the start of the century has been accompanied by steady decline in rainfall. In Guanacaste Province of Costa Rica, associated trends have persisted for 40 years. In India, there are parallel phenomena from a variety of areas. In northwestern Peninsular Malaysia, especially in the Penang and Kedah areas, there has been not only a decline in rainfall, but a less predictable distribution in both time and space, during the past 75 years and particularly during the period since 1960. As a result, some 20 000 hectares of paddy ricefields have been abandoned in this 'rice bowl' of the Peninsula. Another 72 000 hectares have registered a marked drop off in production, leading to an overall shortfall of 27.5% in the Peninsula's rice output in 1978.

Of course it is not (yet) possible to postulate a direct causative linkage between changes in vegetation cover and shifts in rainfall patterns. All manner of other variables are surely at work. Very few problem-specific experiments have been carried out, largely by virtue of the scale and complexity of the processes in question. Nonetheless, in light of the severe repercussions that may arise from the putative linkages, there is urgent need for research on a suitable systematic scale.

Temperate-zone forests: status and prospects

Until recently the expanse of temperate-zone forests has actually been expanding a little through reforestation and afforestation in parts of western

Fig. 2.4. Tree damage of the type associated with acid precipitation (Jim and Julie Bruton, Ardea, London).

Europe and North America. But today certain of the forests are declining, principally (it is supposed) as a consequence of the phenomenon known generically as 'acid rain'. Acid precipitation, as it is more accurately called, appears to derive from a combination of airborne pollutants, working in presumed conjunction with natural stresses such as pathogens and adverse climatic conditions such as drought and frost. It is a disease syndrome induced by multiple factors in association with numerous secondary effects of both biological and non-biological origin. Whatever the precise nature of the phenomenon, it is leading to widespread damage to temperate-zone forests (Fig. 2.4). In Europe alone it is thought to have caused gross injury to almost 200 000 km² of forests, a large proportion of which are dying or are dead (Table 2.2). This total, an aggregate area equivalent to the area of Gt Britain, amounts to some 13% of all forest cover in the region. Similar extensive damage to forests is documented in North America, not only in the more industrialised and populous northeastern part, but in the western United States as well.

Table 2.2. *Acid rain damage to forests in Europe, up to August 1986*

Country	Forest expanse (km²)	Extent of damage* (km²)	Percentage damaged
Finland	194 000	67 900	35.0
Norway	83 330	4100	4.9
Sweden	265 000	10 600	4.0
West Germany	73 230	38 240	51.9
Netherlands	3090	1548	50.1
Belgium	6160	1110	18.0
Luxembourg	820	423	51.6
France	150 750	2796	1.85
Switzerland	12 000	4320	36.0
Austria	375 400	9600	24.2
Italy	63 630	3180	5.0
East Germany	29 000	3500	12.0
Poland	86 770	22 730	26.2**
Czechoslovakia	46 000	12 500	26.1
Hungary	16 700	1837	11.0
Yugoslavia	95 000	10 400	10.9
Totals	1 500 880	194 784	12.9

* Four classes of damage are generally recognised, viz. slight damage, medium to serious damage, critical damage with trees actually dying, and complete damage with trees dead (conforming to needle/leaf loss of 11–25%, 26–60%, 61–90%, and 100%). In West Germany as of late 1985, 32.7% of trees showed slight damage (63% of damaged area), 17% medium to serious damage (32.8%), and 2.2% critical or complete damage (4.2%).
** In late 1986 the Polish Academy of Sciences estimated around 50%.
Extensive forest damage has also been reported in parts of western Soviet Union, but no statistical details are available. Data from Environmental Resources Ltd. in London, Government Ministries of West Germany, and the Economic Commission for Europe.

These depletive processes ostensibly work their effect with remarkable speed. In 1970 there was no injury evident, and as recently as 1980 the scale of the phenomenon was only just emerging. The mechanisms appear to operate in covert fashion for long periods, before exerting their potent cumulative impact. We may well speculate on how far there are similar long-term effects gathering force in other forests of temperate zones, and how much overall damage will come to light within the foreseeable future. So while these depletive patterns are still small as compared with deforestation in the humid tropics, they may well become comparably significant by the end of the century.

In the tropics too, the phenomenon is starting to manifest itself, notably in south-central Thailand, southern Peninsular Malaysia, southern Brazil and southeastern China. In extensive sectors of South America, Africa and southeast Asia, naturally acid soils are likely to prove highly sensitive to the additional ecological burden of acid depositions. In some other areas, such as northern China and parts of India, soils are more alkaline and offer a natural buffering capacity.

Boreal forests: survival outlook

Finally, let us consider the boreal forests, which cover an area of 11.7 million km^2 at high latitudes in the northern hemisphere (Fig. 2.1). This compares with 8.6 million km^2 for tropical forests and 8.2 million km^2 for temperate-zone forests. Boreal forests are threatened by an altogether different type of human agency: the build-up of 'greenhouse gases' and their warming effect on the planetary ecosystem. Let us briefly review the nature and scope of this further phenomenon.

A greenhouse-affected world

By the first quarter of the next century at the latest, and perhaps even by the turn of the century, we shall surely be experiencing the climatic dislocations of a planetary warming, stemming from the build-up of carbon dioxide and other 'greenhouse gases' in the global atmosphere (see Bolin, this volume). Associated with this general warming will be some redistribution of precipitation patterns. The consequences for forests worldwide are likely to be pervasive and profound.

The atmosphere contains about 700 gigatons, or billion tons, of carbon as carbon dioxide (or about 0.03% by volume). Terrestrial vegetation and soils, in forests for the most part, contain at least 2000 gigatons (see Table 2.1), or roughly three times the amount in the atmosphere. The build-up of carbon dioxide in the atmosphere stems primarily from combustion of fossil fuels, which are emitted to the extent of some 5 gigatons a year (see Bolin, this volume). But there is also a sizeable contribution from the burning of tropical forests, as documented by Dr George Woodwell and his colleagues. A realistic figure for this biotic contribution to date is 1.9 gigatons a year, within a range of 0.9–2.7 gigatons. But by the year 2010, the tropical forest source could rise to almost 6 gigatons, as compared with somewhere between 6.2 and 8.4 gigatons for fossil fuels. As the numbers of slash-and-burn farmers continues to increase in tropical forests, the biotic source could contribute as much as 8–10 gigatons a year by 2025 – but then decline sharply thereafter, on the grounds that there will not be many more forests left to burn.

In addition to carbon dioxide, there is a number of trace gases that, molecule for molecule, are far more efficient than carbon dioxide in absorbing infrared radiation from the planetary surface (see Bolin, this volume). These gases include nitrous oxide and methane, some of the main sources of which could well be tropical forests. Other sources include hydrocarbon fuels, farms and oceans, and their contributions are moderately well documented – by contrast with the case for tropical forests, which have been all too little investigated. It is known, however, that tropical forest soils are prodigious producers of nitrous oxide. The gas derives mainly from aerobic micro-organisms, specifically bacteria that oxidise products of organic decay such as ammonia, and decay may well be much more active in disturbed forests and deforested lands.

As for methane, it is thought that major sources are rice paddyfields and ruminants' stomachs, accounting for perhaps one-third and one-fifth, respectively, of the increase in emissions since 1940. Supplementary sources are the burning of vegetation, and termites (which generate significant amounts of methane as they digest cellulose), though the proportionate amounts they account for remain unknown. Termites flourish in tropical forests; and their density is increased by human activities, especially through clearing of the forests to make way for pastures and croplands. The proportion of material consumed by termites in relation to net primary productivity is greatest in wet savannas, which are often created in the wake of tropical deforestation.

What changes can we expect for forest cover in a greenhouse-affected world? We can consider some deterministic projections that reflect just carbon dioxide build-up – bearing in mind that the contribution of trace gases may well be at least as great as that of carbon dioxide. On the grounds of increasing temperatures (moisture and soil factors may have a large bearing too), tropical forests are theoretically projected to expand a little, from 8.6 million km^2 today to almost 10 million km^2 supposing they have not been widely eliminated meantime, and supposing too that there are unoccupied lands into which they can expand. Warm temperate forests may well expand from 21% of all forests today to 25%, and cool temperate forests from 15% to 20%. But the most sweeping changes are projected to occur at high latitudes, where the predicted temperature increase is largest. Thus boreal forests are projected to decline from 23% today to a mere 1%.

While there is some attention given to the impact of changing climate on forests, there is all too little attention given to the reverse process, viz. the influence of changing forests on climate. Forest zones would theoretically 'migrate' polewards in a greenhouse-affected world as global temperatures increase. If climatic changes are sufficiently rapid, trees may not be able to migrate in step due to a marked lack of seeds and other propagules. The poleward edges of forest zones may experience a basic shift in species composition, and degenerate into shrub communities. The outcome would be a marked decline in carbon stocks held in plants and soils, a decline that would in turn release yet more carbon to the atmosphere.

The scale of this further biotic contribution of carbon could be large. So far as we can determine, boreal forests and their soils harbour some 500 gigatons of carbon, or about 25% of the global total. We have only the vaguest idea of how much of this boreal-forest stock could be released into the atmosphere. But as Dr George Woodwell points out, it is not improbable that during the course of several decades it could amount to anywhere between 10% and 50%. A contribution of this order would amount to somewhere between 3 and 10 gigatons a year (to be compared with a projected release of between 6.2 and 8.4 gigatons in the year 2000 from fossil fuels). In short, the boreal-forest release could contribute a substantial increment to the build-up of carbon in the atmosphere.

All in all, we can anticipate that, because of the vital role played by forests in maintaining homeostasis of vegetation/climate interactions, the warming trends of the initial greenhouse effects will continue and even accelerate – and the causative impulse in this case will come from deforestation. In other words, there will be positive feedback effects extending into the indefinite future – or at least for as long as sufficient forest cover persists in the zones in question.

Conclusion: a deforested biosphere?

This, then, is the scope of planetary deforestation that we are engaged in. We shall soon have to start recolouring our maps. Where there have hitherto been broad swathes of green around the globe denoting forest cover, we shall need to employ alternative colours, such as grey or brown, to indicate those extensive territories where forests that have persisted for many millennia, indeed millions of years, have been eliminated within just a single century or so.

Nor will the disappearance of forests be the entire story. A deforested biosphere will be one in which many other vegetation systems will likewise be disrupted through the loss of forests and their capacity to maintain their gyroscopic effects of stability and homeostasis. Essentially we are conducting a global-scale experiment, with irreversible outcome, with many predictable repercussions of impoverishing impact, and with many other unpredictable repercussions whose impact may prove yet more disruptive of biospheric workings that sustain all forms of life. And we are doing it all with scarcely a thought.

Further reading

Bolin, B., Bo Döös, R., Jäger J. & Warrick, R.A., eds. (1986). *The Greenhouse Effect, Climatic Change, and Ecosystems.* Sussex: John Wiley and Sons, Chichester.

Ehrlich, P.R. & Ehrlich, A.H. (1981). *Extinction: The Causes and Consequences of the Disappearance of Species.* New York: Random House.

Environmental Resources Ltd (1985). *Acid Rain: A Review of the Phenomenon in the EEC and Europe* (report prepared for Commission of the Environment Communities). London: Environmental Resources Ltd.

Erwin, T.L. (1983). Tropical forest canopies: the last biotic frontier. *Bulletin of the Entomological Society of America,* spring issue, 14–19.

Farnsworth, N.R. & Soejarto, D.D. (1985). Potential consequences of plant extinction in the United States on the current and future availability of prescription drugs. *Economic Botany,* **39,** 231–40.

Fisher, A.C. (1982). *Economic Analysis and the Extinction of Species.* Berkeley: Department of Agriculture and Resource Economics, University of California.

Food and Agriculture Organization and United Nations Environment Programme (1982). *Tropical Forest Resources.* Nairobi, Kenya: Food and Agriculture Organization of the United Nations, Rome, Italy, and United Nations Environment Programme, Nairobi, Kenya.

Government of Federal Republic of Germany (1986). *1986 Forest Damage Survey.* Bonn: Federal Ministry of Food, Agriculture and Forestry, Government of Federal Republic of Germany.

Henderson-Sellars, A. & Gornitz, V. (1984). Possible climatic impacts of land cover transformations, with particular emphasis on tropical deforestation. *Climatic Change,* **6,** 231–58.

Hutchison, B.A. & Hicks, B.B. eds. (1985). *The Forest–Atmosphere Interaction.* Boston, Mass.: D. Reidel Publishing Co.

MacArthur, R.H. & Wilson, E.D. (1967). *The Theory of Island Biogeography.* New Jersey: Princeton UP.

McCormick, J. (1985). *Acid Earth: The Global Threat to Acid Pollution.* London: Earthscan Ltd.

Moore, P.D. (1985). Forests, Man and Water. *International Journal of Environmental Studies,* **25,** 159–66.

Myers, N. (1983). *A Wealth of Wild Species.* Boulder, Colorado: Westview Press.

Myers, N. (1983). *The Primary Source.* New York and London: W. W. Norton.

Myers, N. (1985). Tropical deforestation and species extinctions: the latest news. *Futures,* **17,** 451–63.

Oldfield, M.L. (1984). *The Value of Conserving Genetic Resources.* Washington D.C.: National Parks Service.

Salati, E. & Vose, P.B. (1984). Amazon Basin: a system in equilibrium. *Science,* **225,** 129–38.

Shukla, J. & Mintz, Y. (1982). Influence of land surface evapotranspiration on the earth's climate, *Science,* **215,** 1498–1501.

Simberloff, D. (1986). Are we on the verge of a mass extinction in tropical rain forests? In D.K. Elliott, ed., *Dynamics of Extinction.* New York and London: pp. 165–80.

Soule, M.E. ed. (1986). *Conservation Biology: The Science of Scarcity and Diversity.* Sunderland, Mass.: Sinauer Associates Inc.

US Department of Energy (1985). *Detecting the Climatic Effects of Increasing Carbon Dioxide.* Washington DC: Carbon Dioxide Research Division, US Department of Energy.

Western, D. & Pearl, M., eds. (1989). *Conservation 2100* (proceedings of conference organized by the New York Zoological Society, late October, 1986) (forthcoming).

Wilson, E.O., ed. (1988). *Biodiversity* (proceedings of National Forum on Biodiversity, Washington D.C., mid-September, 1986). Washington D.C.: National Academy of Sciences Press.

Woodwell, G.M. *et al.* (1983). Global deforestation: contribution to atmospheric carbon dioxide, *Science,* **222,** 1081–6.

Woodwell, G.M. (1986). Global warming, and what we can do about it. *The Amicus Journal* (quarterly publication of the Natural Resources Defence Council, New York), **8,** 8–12.

[3]

Attitudes to animals

Marian Stamp Dawkins

It is all but impossible to make any statement about human attitudes to animals that is both a generalisation and true. There are few areas of our lives where our attitudes to anything are as confused and tangled as they are in the way we think about and treat animals.

This is perhaps not surprising. We are confused enough in our attitudes to just one kind of animal – members of our own species. Just think what humans do to other humans: love them, nurture them, defend them with our lives, kill them, torture them, let them die of neglect, use them as slave labour, admire, revere, despise and so on. Attitudes to other species of animal include all this variety as well as an even greater range of attitudes brought about by the even greater range of roles that animals can play in our lives: they can be food, companions, entertainment, sport, deities or dangerous nuisances. No wonder that our attitudes towards animals vary between extremes.

To quote Albert Schweitzer:

> A man is truly ethical only when he obeys the compulsion to help all life which he is able to assist, and shrinks from injuring anything that lives. He does not ask how far this or that life deserves one's interest as being valuable, nor, beyond that, whether and how far it can appreciate such interest. Life as such is sacred to him. He tears no leaf from a tree, plucks no flower and takes care to crush no insect. If in summer he is working by lamplight, he prefers to keep the window shut and breathe a stuffy atmosphere rather than see one insect after another fall with singed wings upon his table
> (From *Civilisation and Ethics*, translated by C.T. Campion)

Sometimes we find a poetic identification with another species. Here is William Blake:

> Am not I
> A fly like thee
> Or art not thou
> A man like me?
>
> (*Songs of Experience*)

At times this identification can take a truly bizarre form. James Serpell in his recent book *In the Company of Animals* quotes a newspaper report about a Texas hairdresser who was said to be 'somewhat upset' when her pet dog killed and ate her four-week-old daughter. She was, however, nearly hysterical when told that the animal would have to be destroyed. 'I can always have another baby', she is reported to have said, 'but I can't replace my dog Byron.'

Even more interesting is the case cited by Michael W. Fox (*The Case for Animal Experimentation*, 1986) of the New Jersey yachtsman who was charged with the manslaughter of two of his crewmen because he did not throw his labrador dog overboard to make room for them. The yacht had overturned and the two men were floundering around in the water and eventually drowned. There was a lifeboat but there was no room for them because there was a large labrador dog sitting in it. This is of course a particularly interesting example because it is a real-life version of the classical philosopher's dilemma of who you would save in a fire if you could only rescue one person or if there were limited space in a lifeboat. Here was a man who actually had to make such a decision, made it and was consequently accused of manslaughter. He was, incidentally, ultimately acquitted.

These are examples of people identifying closely with animals and behaving towards animals with much the same ethical concern that they would show towards other humans and even, in two bizarre cases, giving them greater ethical consideration than they do other humans. But such attitudes are far from universal. At the other extreme, we find attitudes to animals advocating quite different treatment and putting them in a totally different moral category. Here is a passage from a recent book on wildlife management:

> It would seem obvious that a wildlife manager should try to take the largest harvest possible from an exploited population but the maximum sustainable yield in terms of meat, that is biomass, is not necessarily the most remunerative because meat is not a particularly valuable product. If, for example, an elephant population were to be cropped for meat, the yield would hardly provide sufficient profit to make the enterprise worthwhile. If, on the other hand, the product was ivory, the highest return would be provided by the old males and it would pay to crop this section of the population.
>
> (S.K. Eltringham, *Wildlife Resources and Economic Development*, p. 467)

Some of the language used in scientific journals also conveys an impression of an attitude to animals that again distances animals from humans:

> . . . shivering began within a minute or two and quickly became vigorous and widespread. The next effect was vocalisation. It began with periods of miaowing which became more frequent and of longer duration. Gradually the miaowing changed to growling and yelping. Later, rapid breathing, panting, salivation, piloerection, and ear twitching (were

seen) . . . the cats would suddenly charge blindly ahead, and jump up or cling to the side or roof of the cage, the pupils being maximally dilated. The cats showed compulsive biting: care had to be taken to prevent them biting through the lead of the rectal probe by offering them instead a pencil, on which they would clamp their teeth and eventually gnaw through.
(F. Bergman & W. Feldberg (1978) Effects of propylbenzilycholine mustard on injection into the liquor space of cats. *British Journal of Pharmacology*, **63**, 3–6)

So we are faced with an enormous variety of attitudes to animals, from reverence for life to apparent disregard for whether something is a plant or an animal, from a kind of brotherhood with animals to seeing them as almost like inanimate objects.

In this chapter, I have two aims. First, I want to look at two important reasons that people commonly give for behaving in certain ways towards animals. These could perhaps be said to amount to ethical principles, even though people are sometimes not very explicit about what they are doing. Second, having tried to make sense of the way people respond to animals, I want to argue that our attitudes to animals are in desperate need of facts about them. In other words, our ethical attitudes, even when we can actually identify them, are very likely to be erroneous unless we have considerably greater biological knowledge of the animals concerned than we have at the moment. I shall try to indicate what this knowledge should be.

Now, I would like to emphasise that in saying we need facts to make ethical decisions about animals, I am not saying that moral conclusions arise from facts – I am not about to fall for the naturalistic fallacy. No number of facts can tell you what you ought to do, but facts can and do have a crucial bearing on which actions we believe to be right or wrong. For example, suppose you believed it was morally wrong to eat the flesh of any animal that itself ate other animals. For you, it was an important moral principle that you only ate the flesh of herbivores. Now suppose that you further believed that the only foods that any bird ever ate were seeds and fruit and therefore you could happily eat the flesh of any bird and still live in accordance with your moral principle. Then, however, you begin to discover certain facts about the dietary habit of birds. To your horror, you find that many of them eat earthworms, frogs, insects and all sorts of non-vegetable foods. In the light of these facts you begin to think that you should not eat the flesh of birds and you stop doing so. Now your ethical belief – that it is wrong to eat the flesh of any animal that is not herbivorous – remains absolutely untouched. But the facts – in this case about what birds eat – have radically altered what you believe to be right or wrong for you to eat. And it is in this sense that I am going to emphasise the importance of biological facts about animals, not as a source of moral imperative, but as the guide of moral principles that are already there.

The first things I want to discuss are those moral principles themselves. Are there any in our attitudes to animals? Can we make sense of the extraordinary mixture of attitudes that people have to animals (Fig. 3.1)? I gave some

Fig. 3.1. Different attitudes to animals (lower photograph, *Illustrated London News*).

quotations earlier about different people's attitudes, but we can find almost as much variety within one person. The same person can be quite horrified at eating a cat or dog and yet be quite happy at the idea of eating sheep or pigs. Many British people regard eating cows as quite acceptable but eating horses as barbaric. Animals which look cuddly and appealing, such as baby seals or puppies, evoke more moral concern than animals such as rats or snakes. Animals that are large or rare get more consideration than animals that are smaller and more common. And of course if you are an animal and you want to elicit the greatest human concern, you should combine all these qualities and be large and rare and cuddly and be a Giant Panda.

It is very difficult to see what moral principle someone is operating under if they say on the one hand that their dog must be given the very best food and not allowed to suffer in any way and on the other that a rat or a snake must be killed with very little concern about how. Confusing though this situation often is, it is possible to discern two important ethical strands, two ethical principles, that is, for behaving in certain ways towards some animals and in different ways towards others. Both of these ethical principles have a long history in this connection. Both have been used and are still used by people as their guideline – sometimes explicit but much more often implicit – for how to treat animals. Both would benefit from more factual information.

The first of these moral principles is that people give moral value to animals that are clever or that show evidence of the ability to reason. We are more likely to take ethical notice of an intelligent animal than one which is stupid, as I have often found when talking to people about animal welfare and the problems of modern agriculture and farming methods. When they discover that I work with chickens, they will very often say something like 'Well, they're not very clever, are they?' – as if that let them off the moral hook – as though as long as chickens are stupid, we need not worry about the ethical problems of keeping them in battery cages or broiler houses.

I'm not in any way justifying this view. I'm simply stating it as an empirical fact that it is a view many people hold: cleverness is itself an attribute that deserves moral consideration. (I have a suspicion that this may have something to do with self-interest: clever animals are more likely to find you out and take steps in retaliation if you mistreat them. A stupid animal may not be able to realise the connection between what you do and what happens to it, so you can get away with a lot. But a clever animal may realise what you are up to and either not cooperate or even retaliate. I'm quite sure that if ever we were approached by intelligent beings from outer space, we would immediately give them a great deal of moral consideration because, being intelligent, they would quickly realise if we were up to something and, being powerful, they might well take steps against us if we didn't behave ourselves. But this is by the way. We don't have to justify a moral belief to acknowledge that it is held by a number of people.)

The philosopher René Descartes clearly expressed the view that because non-human animals cannot speak, they couldn't be said to think or reason. And because they couldn't reason, they were in a totally different moral category from that of human beings. He wrote in 1649:

> Speech is the only certain sign of thought hidden in a body. All men use it, however stupid and insane they may be and though they lack tongue and organs of voice; but no animals do.
>
> (Letter to Henry More)

He believed that the absence of a reasoning soul meant that animals did not count for very much ethically either:

> Please note that I am speaking of thought, not of life or sensation. I do not deny life to animals, since I regard it as consisting simply in the heat of the heart; and I do not deny sensation, in so far as it depends on a bodily organ. Thus my opinion is not so much cruel to animals as indulgent to men . . . since it absolves them from the suspicion of crime when they eat or kill animals.

There are two separate arguments here. The first is the assertion that speech is needed to show evidence of thought. And the second is that evidence of thought and reason is itself a basis of ethical consideration. As to the second point – that cleverness is a moral virtue – I've already said that I'm not quite sure where this belief comes from and I'm simply accepting it as a widespread ethical value. But on the first point, this clearly is a case where scientific evidence about animals might have a lot to say. Is it in fact the case that animals that cannot speak have no ability to reason or to think?

The philosopher Stuart Hampshire stated baldly that:

> It would be senseless to attribute to an animal a memory that distinguished the order of events in the past, and it would be senseless to attribute to it an expectation of an order of events in the future. It does not have the concept of order, or any concepts at all.
>
> (*Thought and Action*, 1959)

But one of the most exciting recent developments in the study of animal behaviour has been the realisation of just how clever animals are and that their inability to speak is not so much an indication of an inability to think as an indication that we have to be rather more ingenious in devising ways of discovering whether or not they think.

For example, Herb Terrace and Mark Hood at Columbia University undertook an experiment quite specifically designed to see whether pigeons could develop a concept of order. Pigeons can quite easily be taught to peck a key for the reward of some food: the pigeon is shown a round disc or key which is often illuminated with a coloured light and it very quickly learns the connection between pecking at the key and having some food delivered. What Terrace and Hood did was to present pigeons not with one key but three – a red one, a green one and a blue one. The pigeons had to learn that to get food, they needed to peck the keys in the right order.

Task 1: All pigeons learn to peck 3 simultaneously presented colours
in the order A–B–C.

Then 2 new stimuli, X and Y, are introduced.

Task 2a: For half the pigeons, B is
still in the middle position and
the birds learn to peck in the
order X–B–Y.

Task 2b: For the other half, B is
not in the middle position and
the task is to learn B–X–Y or
X–Y–B.

Learning is significantly faster in 2a than 2b.

Fig. 3.2. Outline of the experiment by Terrace and Hood suggesting
that pigeons can learn the concept of 'in the middle' and carry it over
from one task to another.

The pigeons could learn to do this without difficulty even though the
spatial position of the three keys was altered on each trial. It certainly looked
as though they could recognise that the order in which the keys were pecked
was in fact important. But to be sure that this really was the case, the
experimenters then added a twist. Having trained the pigeons on this first
task, they then required them to learn another task which also involved
pecking three keys in a set order, only this time the green and blue keys were
replaced by two keys, one with a vertical line and one with a diamond
(Fig. 3.2). The red key remained the same. One group of pigeons had, as their
second task, to learn to peck the keys in the order line–red–diamond with the
red key in the middle position just as it had been for their first task. The other
group also had to learn to peck the same three keys in a different order, either
red–line–diamond or line–diamond–red. The point was that, for the second
group, the red key was not in the middle position as it had been in the first
part of the experiment.

What they found was that the pigeons in the first group found their new
task much easier than the pigeons in the second group and learnt much more
quickly to peck the new keys in the right order. Terrace and Hood argue that
the reason for this was that the first group of pigeons had remembered the
fact that the red key had to be pecked in the middle of the series, whereas
the second group had to learn a totally new series. The concept of serial order
with red in the middle was learnt by all the birds in the first part of the
experiment and was a help to the first group and a hindrance to the second.

But important though such studies are, and instrumental though they may
turn out to be in changing people's attitudes to animals, they are certainly
not the only kind of evidence we need. Although the cleverness of an animal
may be important in deciding how to respond to it ethically, there is another,
even more important principle that runs through a great deal of current
thinking about attitudes to animals. This is the view that what matters,
morally, about another organism, human or otherwise, is its sentience, its

capacity to feel pain and to suffer. This view too has a long history and one of its most famous proponents is Jeremy Bentham:

> a full grown horse or dog is beyond comparison a more rational as well as a more conversible animal than an infant of a day or a week, or even a month old. But suppose it were otherwise, what would it avail? The question is not, Can they *reason*? nor Can they *talk*? but Can they *suffer*?
> (*Introduction to the Principles of Morals and Legislation*, 1789)

This quotation has become almost a cliché, which is itself interesting because although it was written 200 years ago, it still obviously strikes a chord in many people. Bentham put his finger on what is the greatest concern to most people about animals – their capacity to suffer. This view is given a great deal of prominence in the writings of Peter Singer, Richard Ryder, Stephen Clark and others who have written on this subject.

This view – that our moral attitudes to animals should be determined by whether or not we believe these animals to have the capacity to suffer is all very well, but it does carry two major problems. The first is: what do we mean by suffering? And the second is: even if we can define it, how on earth would we know that another animal was suffering? Feelings of suffering are subjective, private and inaccessible to any kind of scientific study. This critical question I will come to in a moment and I will argue that this is one of the most important areas in which scientific evidence can help us to make moral judgements.

But first, what do we mean by suffering? What, come to that, do we mean by suffering in humans, let alone non-human animals? In the case of humans, we talk about suffering from hunger, suffering from thirst, suffering from heat, suffering from cold, suffering from boredom, suffering from overwork, suffering from bereavement, and so on. Clearly, when we talk about ourselves, we use the one word to cover a very wide range of different states – with different causes, different physical symptoms and, as we know from our own experience, different subjective states as well. So it is worth asking what all these very different states have in common, because if we cannot say what we mean by suffering in humans, we are not going to have much hope with other species.

About the only feature which everything we might label as 'suffering' has in common is that all such states could be described as 'extremely unpleasant'. They are all states in which people would rather not be and from which they would try to escape if it were at all possible. It is important to emphasise the extremely unpleasant nature of these states because a mild itch which lasted only a short time might be vaguely unpleasant but could hardly be described as 'suffering'. But if someone had an itch that dominated their lives and prevented them from sleeping or doing any work, then we would say that they 'suffered'. So part of the definition of suffering is that the unpleasant emotional state must be unpleasant enough and worrying enough to the organism experiencing it to justify the term. There is also an element of attempting to get out of the situation, highly motivated to do something about the very unpleasant situation it finds itself in. If we say that someone is suffering by being confined in a small space, we mean that they would try to get out of it if they could.

I will enlarge on this idea later on. Before I do that I want to deal with the other difficulty with the use of 'suffering' as a criterion for moral attitudes to animals, namely the logical difficulty of knowing when another organism is suffering.

An objection which is often raised is that we can never know what other animals experience emotionally, and therefore trying to study suffering or trying to use it as a basis for moral action is nonsensical. The main thing to be said to this objection is that it is absolutely right. But then so it is for other people. We can never actually know what other human beings experience. There are exactly the same logical difficulties in knowing when another person is suffering, when, in other words, they are experiencing intensely unpleasant emotional states.

But in the case of other people, we do not let this logical difficulty stand in our way. We are perfectly happy to believe that we can understand what other people are feeling. In other words, we pay lip service to the idea that all subjective experiences are private and inaccessible and then respond to the people around us as though their experiences of pleasure and pain were public knowledge. We then have to raise the question of whether the logical impossibility of attributing subjective experiences to other people, which we quite happily ignore, is really any less than the logical impossibility of knowing what an animal of another species is experiencing. Of course in the case of other people we have words with which we can ask them what they are experiencing, but are words really that important? Surely we can undertand a great deal of what another person is feeling even if we do not speak their language? Counsellors and people who are particularly concerned with understanding the feelings of other people often say you have to 'listen to the music behind the words', you have to listen to what is not being said, you have to read the person's body language. Even philosophers who have put a great deal of emphasis on language as being necessary for reasoning and thought have not, as far as I know, claimed that words were necessary for having feelings. And so, if we believe we can understand the feelings of other people, and words are not the only means we use to do this, perhaps we need not be discouraged by the idea of trying to understand the feelings in non-human species as well.

How might we go about this and what sort of biological evidence should we look for if we want to establish whether or not a given animal has the capacity to suffer and to identify the situations in which it does suffer? There are two kinds of evidence.

1. Physical symptoms: health, absence of disease and injury. If you want to know whether a particular experiment is causing suffering to an animal, or whether a certain agricultural practice causes suffering to the animals concerned, the first thing to find out is whether the physical health of the animals is being affected. Methods of transporting cattle which result in

the animal being bruised, injured, losing weight or dying would, on the basis of this evidence alone, suggest that suffering is being caused. One of the most damning pieces of evidence against the battery cage system for housing laying hens is the fact that the birds often develop brittle bones and weak legs and as a result may end up with fractures that the stronger boned birds housed in more spacious accommodation do not get. Signs of injury and ill-health plus physiological disturbances of their hormonal system are very powerful evidence for suffering, but they may not be enough. An animal may not show signs of physical injury and yet it is still legitimate to ask whether it is suffering, for example, from being confined in a very small cage or kept in social isolation.

For this reason, we may need to turn to a second sort of evidence which may be able to answer the question we really want answered: is the animal experiencing intensely unpleasant emotional states? This evidence is behaviour.

2. Behavioural evidence: behaviour has been used in this way for a long time. Charles Darwin entitled his book on animal behaviour *The Expression of the Emotions in Animals and Man*, in which he advocated the view that behaviour was the outward and visible sign of an internal state of mind in both human and non-human animals. Many people since have agreed with him and have felt that what animals do is the key to what they are feeling. There are many different ways in which behaviour has been used like this. Many different aspects of behaviour have been used to give evidence of animal mind. I want to concentrate on just one of them, the one that I feel comes the closest to telling us whether or not the animal is suffering.

I previously defined suffering in a preliminary way as one of a large number of intensely unpleasant emotional states with the implication that an animal or person in such a state would do their best to get out of this situation. Another way of putting this would be to say that they would be highly motivated to do action that for some reason they cannot perform.

For example, suppose you were confronted with a human being with whom you had no language in common. This person is emaciated, their bones are sticking out and they are grubbing around in the dirt looking for food or desperately begging for food. Would you feel justified in concluding that this person was suffering even though you could not ask them in so many words whether they were? The answer is that of course you would – on the basis firstly, of their physical state and, secondly, from their behaviour: they are obsessed with finding food. Or, in other words, they are highly motivated to find food and yet are unable to find enough to eat.

Of course, it is not very surprising to find that someone who is in danger of starvation also gives top priority to obtaining food, that is, to reducing the risk of starvation, until they become so weak that they cannot even do this. There is obviously a link between the bodily need and the fact that the search for food dominates everything else. Now this link between bodily need and behavioural priority enables us to generalise to non-human animals because they too have recognisable signs of impending death – disease, injury, emaciation – and they too can show evidence of being highly motivated to an act which they are in fact unable to perform.

I would now like to refine the original definition of suffering as the state existing when an otherwise healthy animal is highly motivated to perform behaviour it is unable to perform (perhaps because of physical restraint, perhaps because the right stimuli are not there), or where it is highly motivated to get away from an environment in which it is confined. This is not a universal definition, but it does cover a large proportion of what is generally referred to as suffering.

One of the keys to recognising suffering, then, is the measurement of the animal's motivation to do certain behaviour. Putting the problem in this way enables us to see that animal suffering and animal well-being are not peculiar extras stuck onto the animals as awkward attributes that we sometimes have to consider. It is not that we study biology and look at an animal's behaviour and ecology and then suddenly think, as an afterthought, that we ought to wonder whether it is suffering. On the contrary, suffering is part and parcel of the behavioural priority systems that have been extensively studied by animal behaviour research workers over the last few years.

Animals in the wild have to make decisions throughout their lives about whether, at a given moment, to feed, look out for predators, drink, build a nest, groom themselves and so on. Because many of these activities are incompatible, performing one will preclude the animal from performing the others.

Many studies have now shown that animals base their decision for one or the other behaviour on the risks of death and reproductive failure on the one hand, and opportunities for reproductive success on the other, that the different behavioural options provide. An animal which gave all its attention to feeding, for example, might not die of starvation but it might risk death through being eaten by a predator. Unless its food reserves were so low that it was in imminent danger of starving, it would probably do best to give some of its feeding time over to looking out for danger. Animals from sea slugs to sparrows have evolved complex decision-making systems in which they constantly make such choices about what to do next. They choose between the Devil and the Deep Blue Sea, the Devil of being eaten by a predator and the Deep Blue Sea of starvation. Their choices shift from moment to moment, as danger threatens or recedes, as their own food reserves fluctuate and as light, temperature and other factors in their environment change.

So much is well established in modern biology. I am now arguing that suffering is part and parcel of this highly evolved behavioural priority system that many animals – certainly birds and mammals – have. But suffering occurs in situations, often but not always unnatural ones – in which the normal behavioural priority system has failed for some reason.

For example, an animal in the wild would attempt not to let itself get near a dangerous starvation level because food-finding behaviour would come into play as the reserves became low; the animal would search for and find food and the balance would be restored. However, if the animal were shut in a cage with no food, its food searching behaviour would be to no avail. The animal would become more and more motivated to feed but it is prevented from doing so by the lack of food, and it suffers as a result. This means that if we want to know whether an animal is suffering, one of the most important things we need to know is whether or not it is strongly motivated to perform a behaviour that in fact it cannot do.

The problem as far as our attitudes to animals are concerned is that the environments in which we keep animals are often highly unnatural and so an animal may be highly motivated to do something that now has nothing to do with survival.

For example, if a bird of a species that normally migrates in winter, such as a warbler, is kept in a cage, it will show a great deal of restlessness and repeatedly attempt to get out of the cage at the time when the species normally migrates south in winter. The cage may be quite a large one so that it can fly and stretch its wings, it may be provided with the very best kind of food and be protected from predators and the vagaries of the weather that might well kill it if it were in the outside world. Nevertheless, it will be very highly motivated to escape and to migrate. The evolutionary reason for this is that in the wild, warblers that give high priority to migration in the autumn are the ones that survive and reproduce. In the natural environment, migration is the animal's best way of avoiding starvation over the winter and in the unnatural environment of a cage, this high motivation to migrate still remains. The animal is highly motivated to do something that it cannot do and, because this behaviour totally dominates its life at this time of year, there is a good case for saying that it suffers as a result.

It is, of course, important to know what the real dangers to animals are. It is clearly essential to our assessment of whether or not they are suffering to know whether they are in genuine danger of death through such things as starvation, dehydration, injury and disease. That is why I started by emphasising evidence from physical symptoms. We can often recognise whether there are really risks to the animal's survival by looking at its physical health and well-being. But it is also important to know what the *animal* perceives as the major risks and how it assesses them. In the case of the warbler in the cage, its physical health may be excellent, but it could well be suffering greatly through being highly motivated to do something – migrate in this case – and yet being unable to do so.

The point is that in an unnatural environment such as the ones in which humans commonly keep animals, real risk and risk perceived by the animals may become divorced from each other. What is really dangerous and what the animal perceives as dangerous become separated.

In nature, animals on the whole choose what is best for them. Their behavioural priority systems enable them to avoid that which is most harmful and to seek out and obtain that which is beneficial to them. But in the unnatural environment of a farm or laboratory or zoo, or even a home where pets are kept, the behavioural priority system may still be active, still telling the animal that it must try desperately to do this or to avoid that. Under these circumstances, suffering may result. So if we want to find out whether an animal is suffering, a very great deal obviously hinges on whether we can measure its motivation to do something that in fact it cannot do.

In order to illustrate how this might be done, I would like to take the specific example of work done in Oxford on the motivation of egg-laying hens to various behaviours which they cannot perform in battery cages.

In Britain, over 90% of hens that supply eggs are kept in cages (I am not referring to the hens that are raised to be eaten – none of these are kept in cages). The birds stand permanently on wire floors and have nowhere to perch, to scratch or to dustbathe. They have no nestboxes to lay their eggs in and the wire floors slope towards the front so that as the egg drops from the standing hen, it rolls forward and can be collected automatically.

Are the hens highly motivated to perform behaviour such as scratching in litter, dustbathing and so on which they cannot perform in these cages? They do not die as a result of not being able to dustbathe, so clearly lack of litter in which to scratch and dustbathe does not fall into the same extreme category as a lack of food or a lack of water. However, as I've argued, despite the fact that there seems to be no threat to the health of the animals, it is still possible that they suffer in the sense of being highly motivated. Certainly some of the evidence does seem to suggest that they are indeed highly motivated to dustbathe.

For example, work in Denmark has shown that if hens are kept on wire floors without the opportunity to dustbathe in proper sand or sawdust, they start 'going through the motions' on the bare wire floors. They show all the movements associated with dustbathing even though there is no dust present. They behave as though they are throwing particles into their feathers as they would in real dustbathing, but there is nothing really there. This, it has been argued, shows that they are so highly motivated to dustbathe that they can't stop themselves doing it even when they do not have the right materials.

A second sort of evidence also suggests pent-up motivation. In Denmark, it has also been shown that birds that are deprived of the opportunity of real dustbathing through being kept on wire floors indulge in an orgy of dustbathing when they are subsequently given access to sawdust in which they can have a genuine dustbath. All the hens start dustbathing and dustbathe for much longer than is normally seen in birds kept permanently on litter. This, too, has led people to argue that during the period of deprivation they were very highly motivated.

A third sort of evidence has come from giving hens choices between cages with different sorts of floors. Hens have a clear preference for a floor covered with sawdust over a floor made of wire. If given a choice between two cages of the same size, hens choose the one with sawdust on the floor. They even choose the one with sawdust on the floor if the wire-floored cage is four times as large. Even if the sawdust cage is so small that they cannot turn round when they get into it, they still choose it over a wire cage eight times as large. Nor does it matter what the hens have been used to when they were growing up. Hens reared all their lives on wire floors had just as strong a preference for litter as hens reared on litter.

All this suggests that the motivation which battery hens have to scratch and dustbathe, behaviours that they cannot perform in battery cages, may be quite high, but this evidence has not been universally accepted so I undertook an altogether different sort of experiment.

Biologists often borrow ideas from other fields and one of the most fruitful recent borrowings has been from economics (see Dasgupta, this volume). Economists make a distinction between commodities that show elastic and those that show inelastic demand. Commodities that show inelastic demand are those that people continue to buy whether the price is high or low. Things like bread or, in some countries, rice, show this pattern. They are called necessities and are characterised by the fact that even when they are fairly highly priced, people continue to buy them in much the same amounts: the demand for them does not change very much even when people have to pay a higher price and may have to forgo other things in order to pay for them.

Commodities that show elastic demand, on the other hand, are bought when the price is low, but tend to drop off the shopping list altogether if they become too expensive. People are simply not prepared to buy them if they have to pay too much for them.

Now animals, of course, do not have money and they don't buy things. But they do have something else which is limited, and they can choose to spend it in different ways. Animals have limited amounts of time during each day and they can spend it in various activities such as feeding, sleeping, dustbathing and so on. The 'price' of these different behaviours can be altered by making it more difficult for the animal. For example, I mentioned earlier that it is quite easy to train a bird to peck a coloured key to get food. It is then possible to put up the effective 'price' of food for the pigeon by requiring it to peck not once for each ration of food but twice, four times or even 50 or 100 times. Not surprisingly, when such experiments are done on a variety of animals, food shows inelastic demand: food is so important to animals that they are simply prepared to do virtually anything to get it.

I wondered whether the same might be true in regard to litter on which chickens could scratch and dustbathe. If the chickens' motivation to gain access to litter or sawdust was as strong as the animals' motivation to gain access to food, and if litter, like food, gave rise to inelastic demand, then we might be led to conclude that litter was as important to them as food. In other words, if hens were as strongly motivated to have litter as they were to have food, we might be able to conclude that they suffered without it, in the same way that they suffer without food.

So a further experiment was designed to give hens a choice between a battery cage with food and water, or a cage of the same size as the battery cage, but with litter on the floor. The litter cage had no food or water and the two cages were separated by a short corridor. The hens were tested one at a time and were left with this choice for eight hours and the whole experiment recorded on video-tape. The hens were also given two other tests also

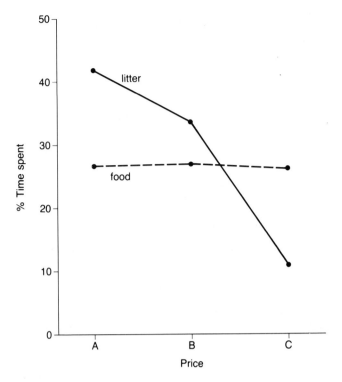

Fig. 3.3. The percentage of an 8-hour period spent by hens in a battery cage with food (broken line) and a cage with litter on the floor but no food. In 'A' tests the hens could move freely between the two by walking along a short corridor. In 'B' tests they had to jump 30 cm to get into either cage, and in 'C' tests they had to push through black plastic curtains.

involving a choice between a battery cage plus food and water or a sawdust-floored cage without. In one test, the birds had to jump about 30 cm from the corridor to get into either of the two cages. In the third test, they had to push through some black plastic curtains. The idea was to change the 'price' of moving out of the corridor into either of the two cages – from no obstacle, to jumping, to pushing through curtains. I then looked at how the hens had allocated their time between the two cages in the face of this price change.

The results were rather startling (Fig. 3.3). Whatever the price, hens spent roughly the same amount of time feeding and ate the same amount of food. The demand curve for feeding showed clear inelastic demand, whatever the hens were required to do for food. But the curve for access to the litter showed a very different shape. When access to the cages was unimpeded, the hens spent more time in the litter cage than in the battery. But if they had to make an effort to move between the cages, they stopped going into the litter: demand for litter appeared to be very elastic.

I repeated the experiment in a slightly different way, this time increasing the 'price' of entering each cage by putting increasing numbers of doors in the way. I used plastic doors like the ones in hospitals so the hens could operate them by simply pushing. The three prices they had to pay were: the two cages side by side with no doors between them (cheapest), the two cages separated by two pairs of doors and 50 cm (intermediate price), and the two cages separated by four pairs of doors and 150 cm (highest price). The result was exactly the same: inelastic demand for food and elastic demand for access to the litter cage.

There are many possible objections to this experiment and I am certainly not saying yet that litter is unimportant to hens. The point is that it is a method of asking hens what is most important to the hens themselves. We are at present engaged in a programme of trying to find out what behaviour is most important for the hens, using the idea of elastic and inelastic demands. We want the hens to rank for themselves which behaviours are most important for them.

This attitude to animals – this way of defining suffering in animals in terms of high motivation to do something which is prevented – is deeply rooted in biology. The animals we deal with in farms, zoos, pet shops, laboratories, were not conceived on a drawing board. Nobody designed them exclusively for human use and convenience. They had wild ancestors, perhaps a very long time ago, and we have only to look at the family dog chasing its ball, burying its bone, turning round to flatten non-existent grass before going to sleep to realise that even in highly domesticated animals there is an evolutionary legacy from the past. Behaviour that helped their ancestors to survive is still with them, even though it may now have no bearing on their health or ability to reproduce.

The capacity of animals to suffer and to feel pain has, or at least had in the past, a biological function just as much as their colour, the length of their legs or their mating patterns. In an evolutionary sense, animals suffer because they have complex decision-making processes which enable them to avoid the major risks of wild animals – starvation, being eaten by another animal, and so on. Suffering is the prolonged activation of the risk-avoidance mechanisms where the animal is unable to carry out the normal risk-avoiding behaviour, like a fire-alarm going off and no-one being able to do anything about it.

Fear, for example, is part of a wild animal's mechanism for avoiding predation and evading or escaping from situations in which the animal is likely to be killed or injured. If the animal is unable to take the steps for avoiding these dangers that it would be able to in the wild, such as running or flying away, it suffers through being highly motivated and being unable to perform the escape behaviour.

It is no use saying to an animal: 'You must be all right because you have food and water and are warm and I have vaccinated you against every known disease', because the animal, trapped in a small cage and peered at by humans, even very well-meaning humans, may believe it is in imminent danger of being eaten by a predator. Its entire motivational decision-making mechanism might be telling it to escape at all costs because *in nature* to be trapped and unable to escape within a few inches of a large animal with staring eyes, has for millions of years meant almost certain death.

We must be prepared to acknowledge that animals in situations that are unnatural, by which I mean different from the environment in which their ancestors evolved, can be highly motivated to do something they cannot do, even though they appear to us to be perfectly healthy and in no danger at all. This is why the measurement of motivation, particularly of frustrated motivation and the way an animal expresses it, is so important. This is an example of what I meant when I said at the beginning that we need biological knowledge about the animals to help us with our ethical judgements about how they ought to be treated.

This leads me on to another very important point, which is really the basis of everything I have tried to say about animals. It is that we must begin to take much more notice of the animals' point of view. I'll go even further. I believe that one of the most important questions in deciding what our attitudes to animals should be is whether we are dealing with a being that actually cares what we do to it. It might be considered immoral to knock down York Minster on many different grounds – that it is beautiful, that it is unique, that it means a great deal to the people who worship there. But York Minster itself would not care. You might argue that it would also be immoral to hack a large whale to pieces on many of the same grounds – that it is a magnificent animal, that it is irreplaceable, that it is a member of an endangered species, that it is unique and so on. But in addition, the whale itself would care what you did to it. It would have a point of view. And in answering the question about what our attitude to animals should be, what we need most urgently to know is whether they have a point of view and what that point of view is.

There is a BBC Radio 4 programme with the brilliant title 'Does he take sugar?' In four words, this title manages to sum up much that is wrong with the attitudes of even very well-meaning people to the disabled, namely that they do not take into account the opinion of the most important person concerned. We are guilty of something very similar with respect to animals. If we are arguing that they ought to be treated in certain ways because they have the capacity to suffer (they are beings to whom things matter), then we should at least take the trouble to ask them what their attitudes are and what matters to them. Only in this way will we arrive at a balanced evaluation of the effects of what we are doing to them. I have tried to give some indication of how we might go about doing this, at least in some circumstances. The methods that I described for doing this involved making use of the animals' own inbuilt decision-making abilities, their capacity to recognise risk and to take steps to avoid it using the emergency procedures built into them by natural selection.

From this, it would seem to follow that we would find the clearest evidence of suffering in animals that have the capacity and the wit to do something about unpleasant situations in which they may find themselves. This in turn may signal a convergence between being concerned about animals that are clever and being concerned about animals that can suffer. I introduced these ideas with the implication that the Cartesian idea of taking ethical notice of the ability to think and reason and the Bentham idea of taking ethical notice of the capacity to suffer were quite separate. Indeed, the quotation I gave you from Bentham, implied that speech, reason and the capacity to suffer were totally distinct. However, I have now argued that one of the best ways to discover whether an animal is suffering is by seeing whether it will work and make an effort to get out of situations that are unpleasant to it. This of course assumes that it has at least the bare minimum of intelligence to work out how to do so. So clearly the methods I am proposing favour intelligent animals because they are the ones that will be able to do this.

It is most likely that the capacity to suffer has evolved in animals with the ability and intelligence to do something about their situations. A tree which is having its branches cut off is unlikely to suffer in the way an animal would from having its limbs cut off because the tree cannot do anything about its situation. If it could flip its branches and shake off the attacker, the sensations of pain might serve the purpose of telling it when to act. But there would be no evolutionary point in a tree that stood there suffering in silence. It is the doers that suffer, the active changers of their environment – animals. The capacity to suffer is part of their armour for avoidance and escape. So it may well turn out that it will be the clever animals that have the most highly developed capacity to suffer. People have an intuitive sense that clever animals deserve more moral consideration than stupid ones. When we know more about the animals themselves, this view may be justified because clever animals may have the greatest capacity for suffering. The distinction I started out with may disappear in the future.

In conclusion, I have tried to argue that our attitudes to animals are muddled, confused and often ill thought-out. I have picked out two of the most prevalent – the view that we should give moral consideration to animals that are clever and the view that we should give moral consideration to the ones that can suffer, although I have also argued that these may turn out to be less distinct than they appear at first.

I have deliberately not tackled other attitudes to animals, such as the idea that we should have different attitudes to them depending on whether they are rare or not, whether they are beautiful or not or whether they are harmful or beneficial to humans. These attitudes are important, but we can also see them in our attitudes to York Minster, Stradivarius violins or the maintenance of rare species of plants because they might be useful one day for medicinal purposes (see Myers, this volume).

I have concentrated instead on animals as organisms that may care what is done to them. If it matters to the animal itself what is done to it, a whole new set of moral considerations comes into play. I have argued that what we need are the animal's point of view, the animal's attitudes and more biological facts about animals, particularly about their behaviour and their expressed preferences for certain situations over others.

I would just like to clarify one further point about what I have been saying in respect of the study of suffering in animals. I have, in a quite brazen way, talked about subjective feelings, for example about the unpleasant feelings we refer to as 'suffering'. I did acknowledge that there might be philosophical problems in trying to make any objective study of the private experiences of other beings, and then carried on as though there was no problem. So did I really mean to leave the impression that we can study the private subjective experiences of other animals without resort to analogy with ourselves? The answer is that of course we cannot. Subjective feelings are private, known only to the organism experiencing them. To know what another organism, even another person, is experiencing, we have to make an analogical leap, an assumption that what they are experiencing is something like that which we are experiencing.

But there is a very important distinction between the different ways we might use such analogies, that I am going to call 'broad-bridge' and 'narrow-bridge' analogies. Take Thomas Nagel's question: 'What is it like to be a bat?' If we make a broad-bridge analogy, we assume that being a bat is just like being a human inside a bat skin. So if you dressed up in a bat suit and flapped around a bit, you would know what a bat felt. But if I asked you, as the human being inside the bat suit, what you would feel about hanging upside down for long periods of time and having nothing but large hairy insects to eat, you might well say that you would find this very unpleasant. You would say this if you were making a broad-bridge analogy and dumping your human feelings wholesale onto the bat.

But suppose you have more information about bats. You knew that, even given a very wide choice of alternative foods, bats choose to eat large hairy insects in preference to anything else. Even when free in the countryside, they choose to sleep hanging upside down. Then you could make a rather narrower-bridge analogy. Instead of assuming that the bat experienced exactly the same thing as you, as a human would if you ate hairy insects, you assume that the analogy is with your feelings when you eat your favourite food. There is still an analogy, but now it is to assume that just as you get pleasure from eating your favourite food, so does the bat. Biological facts about bats would have tempered and narrowed your analogy.

The same is true, of course, in our dealings with other people. Understanding the feelings of other people is not easy. Someone who says 'I know exactly how you feel; I've been through the same thing myself' will probably not understand you nearly as well as somebody who takes the trouble to ask you what you are going through. With other people, we have words. With other species of animals, we have to ask what they are going through in more subtle ways. I have suggested that facts about animal motivation and their preferences for some environments over others are the non-verbal equivalents.

The only way we can arrive at a moral position in our attitudes to animals is to find out more about those animals themselves and their view of the world. In other words, in order to decide what our attitudes *to* animals should be, we need to find out what the attitudes *of* those animals are. We need the humility to admit that it is not just human beings that have attitudes, and that non-human animals may have their own attitudes too.

Further reading

Dawkins, M.S. (1980). *Animal Suffering: the Science of Animal Welfare*. London. Chapman and Hall.

Serpell, J. (1986). *In the Company of Animals*. Oxford: Blackwell.

Singer, P. (ed.) (1986). *In Defence of Animals*. Oxford: Blackwell.

[4]

How many species?

Robert May

Introduction

Charles Darwin provided the essential elements of the explanation for how species originated and thus how life has evolved on earth. This work has changed forever the way educated people see themselves in relation to the rest of the natural world.

Although correct in essentials, Darwin's ideas had some major technical problems in their own time. For one thing, in the absence of the then-undiscovered nuclear forces, it can be shown that neither the sun nor (by a separate argument) the earth can be more than a few tens of millions of years old. For another thing, heritable variation is roughly halved in each generation if inheritance blends the characteristics of mother and father (as was thought to be the case in Darwin's day), making it hard to understand how such variability – the raw stuff on which selection can act – is maintained. A widespread recognition that genetic inheritance operates in a discrete, particulate way, tending to conserve variability, had to await the rediscovery of Mendel's work some 50 years later. Many other questions, including the mode and tempo of evolutionary change, the role of 'neutral selection' as gene frequencies drift under random statistical fluctuations, the selective advantage of sex, and other topics, remain active areas of research today. But all this work takes place within the sturdy framework erected by Darwin.

Given a basic understanding of how species originate, the next question would seem to be how we use this understanding to estimate – from first principles – how many species are likely to be found in a given region. Darwin, who often illuminated his writings with vivid images, conjured up a barrel into which wedges of different sizes and shapes were being hammered, as a metaphor for some physical environment wherein a variety of different species contended for co-occurrence. But Darwin never pursued the question of how many wedges of a given kind (birds, flowering shrubs, and so on) we might expect to find in a given barrel.

I regard it as rather surprising that this question of 'how many species?' has received relatively little systematic attention, from Darwin's time to our own. At the descriptive level, we do have some systematically compiled data and some phenomenological rules deduced from them. Conspicuous among this work are the relations between the number of species and the area of real or virtual (mountain tops, etc.) islands they inhabit, as surveyed by MacArthur and Wilson in their influential book on island biogeography. There is also an expanding literature on empirical patterns of commonness and rarity, often expressed in terms of the relative abundance of species of a given kind (birds, moths, beetles, grasses, and so on) within specified habitats; some of this work is discussed further below. But these empirical patterns represent only a first step, and the second step of explaining the patterns in terms of some fundamental understanding of how species interact with each other and with their physical environment is largely lacking.

In part, such lack of fundamental understanding derives from the sheer difficulty of the enterprise. To understand fully how many species of a given kind are likely to be found in a specified woodlot or sand dune, we would need to answer, at very least, the following nested hierarchy of questions: how each individual population is regulated (so that births balance deaths on broad average), and how this population-level regulation ultimately derives from the behaviour of individual organisms; how species interact as competitors, mutualists, or prey and predator, and how these interactions affect the likelihood of their persistent co-occurrence; how the addition or deletion of particular species affects the continued existence of other species (that is, how food web structure affects community dynamics); and how changes in the physical environment affect the organisms at every level from individual behaviour to ecosystem structure. This is a daunting set of problems, and it is perhaps not so surprising that little progress has been made.

I hope that the reader will not be disappointed that what follows is not an outline of the answers to the above questions. It is not even a systematic review of the current state of play. Rather, I focus on a few particular areas where interesting developments are currently taking place. The motivating theme, to which I will return at the end, is to understand how many species there are, and why there are not more or fewer.

Identifying what regulates individual populations

There is obviously little hope of deducing why there are roughly 700 species of birds that breed in North America, rather than 7 or 70 000 species, if we cannot identify the factors governing the abundance of any one species. Yet, after more than a century of systematic research devoted to determining what, if anything, regulates particular natural populations, ecologists still quarrel over basic issues.

Given that all species appear to have the innate capacity to increase from generation to generation, the task is to untangle the environmental and biological factors that hold this intrinsic capacity for population growth in check over the long run. The task is made more difficult by the variety of dynamical patterns exhibited by different populations. Some populations remain roughly constant from year to year, others exhibit regular cycles of abundance and scarcity, while yet others vary wildly, with outbreaks and crashes that in some cases are plainly correlated with the weather, and in other cases are not.

In an attempt to impose some order on this bewildering variety of patterns, one school of thought sees the relatively steady populations as having 'density-dependent' growth parameters (with rates of birth, death, and migration depending strongly on population density), while the highly varying populations have 'density-independent' growth parameters (with vital rates fluctuating in response to environmental events, wholly independent of population density). This dichotomy has its uses, but can lead to difficulties if it is taken too literally. For one thing, no population can be driven entirely by density-independent factors all the time. No matter how severely or unpredictably birth, death, and migration rates may be fluctuating around their long-term averages, if there were no density-dependent effects the population would, in the long run, either increase or decrease without bound (barring a miracle by which average gains and losses exactly cancelled). To put it another way, it may be that on average 99% of all deaths in a population arise from density-independent causes, and only 1% from factors varying with density. The factors making up the 1% may seem unimportant, and their cause may be correspondingly hard to pin down. Yet, whether recognised or not, these factors will usually determine the long-term average population density.

We may think of the density-dependent effects as a 'signal' tending to make the population increase from relatively low values or decrease from relatively high ones, while the density-independent effects act to produce 'noise' in the population dynamics. Our task is to separate the signal from the noise. For populations that remain relatively steady, or that oscillate in repeated cycles, the signal may be fairly easily characterised (even though the causative biological mechanism may remain unknown). But for irregularly fluctuating populations we are likely to have too few observations to have any hope of extracting the signal from the overwhelming noise.

Two further complications have come to be recognised only in the past few years.

First, even if there were some purely deterministic equation – all signal and no noise – that predicted future population size reliably, the non-linearities inherent in even the simplest such density-dependent equation can produce bizarre dynamics that may indeed have the appearance of being random noise! Consider, for example, a population that has discrete, non-overlapping generations, as do many temperate zone insects, and whose population dynamics we can describe by the relation:

$$N_{t+1} = aN_t(1-N_t),$$

where N_t is the density (normalised to be less than unity) in generation t, and

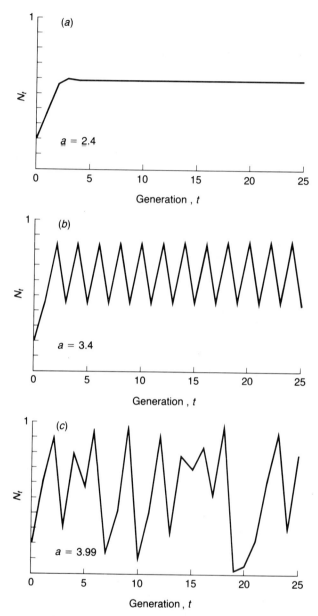

Fig. 4.1. Examples of numerical changes in hypothetical populations with discrete, non-overlapping generations, whose growth rate is described by the equation

$$N_{t+1} = a\, N_t\, (1 - N_t).$$

The size of the population in a given generation, N_t, is plotted against generation, t, for three different values of a: (a) = 2.4; the population settles to a steady value; (b) = 3.4; the population shows steady cycles, repeating every two generations; (c) = 3.99; the population size wanders, apparently at random.

N_{t+1} is the corresponding density in generation $t+1$; a is some defined constant for a particular system.

This equation can be iterated (on a hand calculator, for example) to simulate population changes over many generations, and the successive values of N_t plotted. Strikingly different results are obtained for different values of the constant a (Fig. 4.1). If a has a value greater than 1 but less than 3, then the population settles to a steady value, as our intuition suggests. If a is greater than 3 but less than 3.570, then the population settles into a steady cycle, alternating between high and low values (and repeating every two generations for a at the low end of the range, or every 4, 8, 16, . . ., 2^n generations as a increases). For values of a between 3.570 and 4, this simple and purely deterministic equation describes an apparently random or 'chaotic' population trajectory. (For values of a greater than 4, N runs away to minus infinity (i.e. extinction) in this excessively simple model.) Weird as this spectrum of behaviour may be, it is not peculiar to this equation. Rather, it is generic to essentially all difference equations describing a population with a propensity to increase at low values and to decrease at high values. Similar behaviour arises if there are many discrete but overlapping generations, or even if the population growth is continuous but has time delays in the regulatory mechanisms.

The mathematical properties of these 'deterministically random' or 'chaotic' phenomena were first set out by Myrberg in 1962, and subsequently rediscovered independently by several people. But they remained a relatively arcane mathematical curiosity until they were independently rediscovered yet again by population biologists such as Yorke, Oster and myself in the early 1970s. Since then, the subject has grown explosively, with deterministic chaos finding applications in fluid turbulence, circuit theory, structural mechanics, plasma physics, and elsewhere. In retrospect, it seems odd that such chaotic dynamics were not noted earlier, because entomologists and fisheries people studied equations like the one above as early as the 1940s and 1950s. Although these workers did find chaotic and cyclic dynamics in their numerical studies (carried out on mechanical calculators), they were looking for stable solutions and, having found them, they went no further. It is interesting that the simplest possible mathematical model for a host population with discrete generations, regulated by a lethal pathogen that spreads in epidemic fashion through each generation before reproduction, has *only* chaotic solutions – no stable points and no stable cycles. Deterministic chaos might have forced itself on our attention long ago if only someone had thought to look at such a model.

The lesson to be drawn from these simple models is that density dependence can give rise to a wide range of dynamical behaviour, from constancy, through stable cycles, to apparent chaos or randomness, even in the complete absence of any noise. It is therefore not surprising that many regularly oscillating populations are found in nature. More confusingly, we have the possibility that some irregularly fluctuating populations may be driven by non-linear signals (corresponding to strong density dependence), and not necessarily by density-independent environmental noise. Thus there is considerable irony in the classical disputes between Nicholson (who believed that density-dependent effects were pervasive, holding most populations relatively steady) and Andrewartha and Birch (who held that density-independent effects were generally the rule, causing most populations to fluctuate markedly) when we observe that strong density-dependence will typically produce erratic fluctuations!

A second complication in the extraction of signal from noise in population data can arise when the organisms are distributed unevenly in space, as most are. Most conventional techniques for analysing data make the assumption (often implicitly) that organisms are distributed homogeneously, or at least that one can deal meaningfully with overall averages of the population density. For instance, a conventional technique called 'K-factor analysis' compares overall population densities at each life stage with the corresponding average densities in the next generation, with the aim of determining which stages in complex life cycles account for the density-dependent signals. But if the overall population is distributed non-uniformly over many patches, with different densities in different patches, it can be that much of the density-dependent regulation takes place within patches. Hassell has shown that, in such circumstances, the density-dependent signals can be effectively masked by noise if one seeks them by conventional comparisons of overall average densities in successive generations. It could be that new techniques might overcome these apparent difficulties, but equally it could be that elucidation of density-dependent mechanisms may often require patch-by-patch studies, at a level of detail that is finer than that found in most existing field work.

In summary, it seems clear that all populations are regulated by a mixture of density-dependent and density-independent effects, in varying proportions. The interesting questions are how life-history parameters (having to do with birth, death, and movement) evolve under such different circumstances, and what is the consequent balance between non-linear signal and noise. I believe that our growing awareness of the technical complexity of these issues will lead to better design and more incisive analysis of field and laboratory studies, and thus to a better understanding of factors regulating the abundance of specific populations.

Beyond this, the next steps are to understand interactions among populations, and thence the structure and function of food webs and ecosystems. Rather than dwell on tentative advances in these areas, I will discuss some broad patterns in the distribution and abundance of individual organisms and of species.

Relative abundances of species

It is obviously of fundamental importance to understand why some species are common and some are rare. Such understanding has important applications in conservation biology and elsewhere. Beginning with the

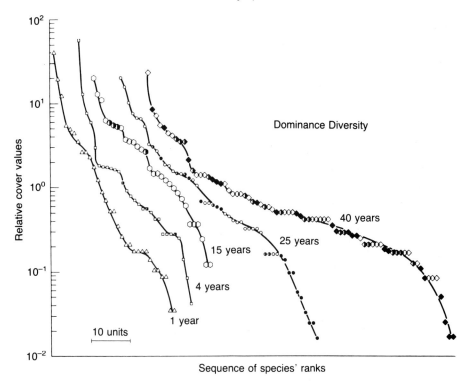

Fig. 4.2. In early successional communities, a handful of species dominate, while in later stages the distribution of relative abundance tends to become more even.

Patterns of species' relative abundance in old fields of five different stages of abandonment in Southern Illinois were studied by Bazzaz. For each stage, the species present are ranked from the most to the least abundant and the percentage that each contributes to the total area covered by all the species in a community (relative cover) is plotted against the species' rank. The successive curves are staggered to the right to avoid confusing overlap. '10 units' corresponds to 10 species ranks. Symbols are open for herbs, half-open for shrubs, and closed for trees.

pioneering work of Robert MacArthur and others, a good deal has been done both to elucidate the variety of patterns of species' relative abundance found in nature, and to study the mechanisms producing these patterns.

In early successional communities, and in environments disturbed by toxins or 'enriched' by pollution, steeply graded distributions of relative abundance are commonly seen, with a handful of dominant species accounting for most of the individuals present. Conversely, in relatively undisturbed 'climax' communities consisting of many species, fairly even distributions of relative abundance are typical; very often, such patterns of relative abundance are distributed according to a so-called 'canonical log normal' distribution, which is discussed below. As illustrated in Fig. 4.2, such

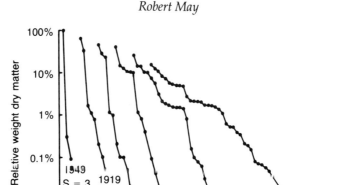

Fig. 4.3. Severe disturbance sets in train a 'reverse succession' from quite even distributions of relative abundance to dominance by a few species.

Changes in the patterns of relative abundance of species in an experimental plot of permanent pasture in the Parkgrass Study at Rothamsted, following continuous application of nitrogen fertilisers since 1856. The patterns are expressed as in Fig. 4.2, except that relative abundance is expressed in terms of each species' contribution to the total weight of dry matter present. Species with abundance less than 0.01% were recorded as 0.01%.

Notice that time runs from right to left; the patterns look like the successional patterns of Fig. 4.2, running backwards in time.

trends in relative abundance of species from dominance to evenness tend to show up in studies of succession of species when abandoned farmland reverts to unmanaged woodland. The effects of pollution or other systematic and sustained disturbances reveal the same trends, except that time effectively runs backward, so that the trend is from evenness to dominance. This kind of 'reverse succession' pattern of changing relative abundance of species under severe disturbance is illustrated in Fig. 4.3. This diagram shows the relative abundance of grass species in grass plots that were disturbed (beginning in 1856) by sustained and heavy application of one particular fertiliser or nutrient, in an experimental study at the Rothamsted Research Station.

The relative abundances within a fairly large group of species very often follow a distribution called 'log normal' as illustrated in Fig. 4.4. This is not surprising as the relative abundances are likely to be governed by the interplay of many more or less independent factors and it is in the nature of the equations of population dynamics that these several factors should compound multiplicatively. It is well known to statisticians that, according to the Central Limit Theorem, such a product of factors leads to a log normal distribution. That is, the log normal distribution arises from products of random variables, and factors that influence large heterogeneous assemblies

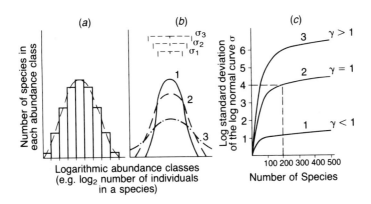

Fig. 4.4. The log normal distribution.
(*a*) A plot of the number of species falling into logarithmic abundance classes in a community of S species (log$_2$ number of individuals is commonly used and represents a doubling of individuals in successive classes). Such data from real communities often conform to a symmetrical, *normal* distribution on the log axis (dashed line). Distributions of this sort are said to be *log normal*.
(*b*) Three curves showing three of the infinite number of log normal cuves which could represent the relative abundances of S species. Curve 1 has fewer rare and fewer very abundant species than curve 3, and hence a narrower spread, here represented by one standard deviation, σ, either side of the mean.
(*c*) S, the number of species in the community, and σ, the standard deviation of the log abundances of the constituent species, are linked by the parameter γ. The *canonical log normal* distribution for S species is the special case among all the possible log normal distributions in which γ = 1, e.g. for 200 species, the canonical log normal distribution of the relative abundances has a standard deviation of 4.05 log units.

of species indeed tend to do so in this fashion. This general observation, however, tells us nothing about the relationship between σ (the standard deviation of the logarithms of the relative abundances) and S (the total number of species present). The remarkable and puzzling fact is that very many assemblies of particular groups of organisms – birds, moths, gastropod snails, plants, diatoms, and others – have patterns of relative abundance of species that obey the 'canonical log normal' distribution. That is, the distributions have the unique relationship between σ and S, illustrated by the curve labelled γ=1 in Fig.4.4, even though this curve represents only one of an infinite family of possible log normal distributions.

In 1975, I suggested that the canonical property may be merely an approximate mathematical property of all log normal distributions when large numbers of species are present (large S). The parameter γ can be estimated if S and the total number of individuals, N, are both known. By making plausible assumptions about the likely range of the ratio N:S, I concluded that γ was unlikely to be less than about 0.2 or greater than about 1.8. I thought this range of γ-values could encompass the data in a reasonable way. The data put together by George Sugihara, and shown in Fig. 4.5, make it clear, however, that real distributions of species' relative abundance obey the canonical relation (i.e. γ=1) more closely than can be explained by such mathematical generalities alone.

Why should this be so? Sugihara has suggested a biological mechanism that will produce the observed patterns. He begins with Hutchinson's observation that individuals and populations inhabit a kind of multidimensional 'niche space', whose n dimensions are those physical and biological factors that affect survival and reproduction (temperature, humidity, availability of nest sites, and so on). Sugihara then imagines that the community inhabits some defined volume (or 'hypervolume') in this *n*-dimensional space, and that this hypervolume is broken up sequentially by the component species in such a way that each of the S fragments denotes the relative abundance of a species. This pattern of sequential breakage (in which any fragment is equally likely to be chosen for the next breakage, regardless of size) seems not inconsistent with evolutionary processes. The solid line in Fig. 4.5 shows the average relation between σ and S predicted by Sugihara's model, and the error bars show the range of plus and minus two standard deviations about this average. The line approximates to the canonical relationship (γ=1) and fits the observed data very well. The fit of the model to the observed distribution patterns does not prove it is necessarily correct; it is always possible that other biological assumptions could produce similar distributions in the relative abundance of species.

In short, there are regularities in the way individual organisms are distributed among species, and even some tentative understanding of how such patterns may arise. But much remains to be done.

Number of individuals versus physical size

There are many other patterns in the distribution and abundance of organisms, most of which have received little attention. For example, in a given region, what is the relation between numbers of individuals and their physical size (mass or characteristic length)? How is the number of individual animals in the 0.1–1.0 cm size class related to the number in the 1–10 cm class?

A set of British authors (Morse, Lawton, Dodson and Williamson) have compiled some facts bearing on this question for insect populations. They have also advanced a qualitative explanation for these facts. They begin with the assumption that roughly equal amounts of energy flow through each size category; although very unlikely to be true in general, this assumption is supported by some evidence from terrestrial organisms ranging widely in size. If the energy typically required to sustain life was the same, gram for gram, for organisms of all sizes, then we would expect the number of

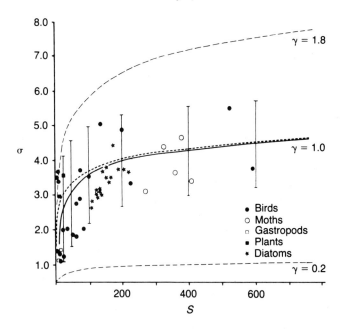

Fig. 4.5. Real distributions of species abundance are closer to the canonical relation ($\gamma = 1$) than can be explained by mathematical generalities alone (which predict a value of γ between 0.2 and 1.8 for realistic ratios of the number of individuals and the number of species in a community).

A plot of σ, the standard deviation of the logarithms of the relative abundances, versus S, the number of species, for various communities of birds, moths, gastropod snails, plants and diatoms. The solid line is the average relation predicted by Sugihara's model of sequential niche breakage, and the error bars represent plus and minus two standard deviations about this mean.

individuals, N, in the size class with characteristic mass, M, to be related to M (i.e. will scale with M) according to the simple relation $N \sim M^{-1}$, as the (assumed constant) energy was divided into a larger number of smaller parcels in the smaller size classes. Characteristic mass is related to characteristic length, L, by the geometric scaling relation $M \sim L^3$, since mass rises with volume. So the simple expectation $N \sim M^{-1}$ translates into the relation $N \sim L^{-3}$. But, in fact, metabolic costs tend to become relatively larger at smaller sizes, partly because smaller creatures have higher rates of surface area to volume than do bigger ones; smaller creatures tend to have higher heartbeats, and, as it were, generally to 'live faster' and die sooner. It turns out (and I omit all details here) that this translates into an expectation that N will scale with M according to $N \sim M^{-3/4}$. Combining this with the geometric relation $M \sim L^3$, we thus arrive at the expectation that the total number of individuals, N, in the size class with characteristic mass M and length L may scale with body size as $N \sim M^{-3/4} \sim L^{-9/4}$. That is, for a ten-fold decrease in characteristic length

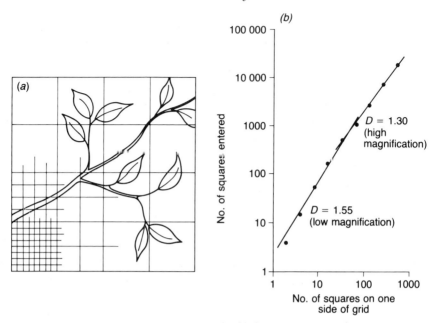

Fig. 4.6. The perceived edge length of habitats increases as the step-length of measurement, λ, decreases. The relationship between them involves a 'fractal dimension', D, which can be estimated by the procedure illustrated.

(*a*) Photographs of plants at various magnifications were placed under a grid by Lawton and co-workers. The number of squares entered by the outline of the plant were counted, starting with a coarse grid of two large squares on one side, then 2^n squares, with n varying from 2 to 6 or 7, depending on the grid size. For ease of representation, the plant's leaves in this figure are drawn flat; in reality they are oriented at all angles with respect to the grid. Also for clarity, the progressively finer divisions are illustrated only in one corner of the diagram. The logarithm of the number of squares entered by the outline of the plant is then plotted against the logarithm of the number of squares along one side of the grid, as shown in (*b*). The slope of the line equals the fractal dimension, D.

(*b*) Data gathered in this way for Virginia Creeper, photographed without leaves in early spring. The twigs were photographed at one scale, then parts of the same twigs were rephotographed at a higher magnification, permitting D to be estimated at two levels of resolution. D may then be used to predict how the perceived length of the outline will change with changes in the step-length of measurement, and therefore with the body size of the animals using the habitat.

we would, on this basis, expect a roughly 180-fold increase in the total number of individuals ($10^{2.25}$ is roughly equal to 180).

As seen through the eyes of individual organisms, however, the structure of the habitat – and hence the number of possible ways of making a living – is unlikely to scale linearly with L. Lawton and co-workers pursue this complication by using recently-developed ideas about the 'fractal geometries' found in nature. Consider, by way of illustration, the length of the coastline of Britain. If we measure it on a 1-km scale we get one answer. Measuring on a 10-m scale, we would get another, larger answer. A yet larger answer would be obtained on a 1-cm scale, and so on. The coastline of Britain is thus not simply one-dimensional, but has 'fractal dimension', D, such that the perceived length depends on the step-length of measurement, λ, as λ^{1-D}. If $D=1.5$, for example, a ten-fold reduction in the measurement scale (from, say, 1 m to 10 cm) will result in the apparent length increasing by a factor of $10^{0.5}$, or roughly 3. Lawton and his colleagues applied these notions to measure the profiles of various kinds of vegetation at different scales, concluding that D in the habitats of interest to them ranged from around 1.3 to around 1.8, with an average value of around 1.5. The operational procedure by which these estimates were arrived at, from studies of real plants, is illustrated in Fig. 4.6.

This all means that, for herbivorous insects that exploit their surroundings in an essentially one-dimensional way (using the edges of leaves, or the like), a ten-fold decrease in physical size produces a roughly three-fold increase in the apparently available habitat. For creatures exploiting their environment in an essentially two-dimensional way (using surfaces rather than edges), the effect must be squared, so that a ten-fold decrease in physical size produces an effectively ten-fold increase in apparent habitat. These two factors – the one-dimensional factor 3 and the two-dimensional factor 10 – are likely to put lower and upper bounds on the range of possibilities found in actual assemblies of insects.

Combining these fractal aspects of habitat perception with the metabolic considerations summarised in the equation above, Lawton and co-workers conclude that a ten-fold decrease in characteristic length, L, is likely to produce an increase in N, the total number of individuals in a size class, of between 500 and 2000 (that is, roughly between three and ten times 180). As illustrated in Fig. 4.7, this very rough expectation is borne out surprisingly well by data for the number of individual arthropods of different body lengths found on vegetation in places ranging from primary forests, primary riparian vegetation, and secondary vegetation in the New World tropics, to temperate habitats such as birch trees on Skipwith Common in North Yorkshire. It is easy to fault the crude assumptions and approximate estimates inherent in this work, but I think the study points the way to important new directions for empirical and theoretical research.

Number of species versus physical size

Other patterns can be sought in the number of species in different categories of physical size, within a given region.

Fig. 4.8 presents one of the few systematic studies of this question, showing the way in which all 3000 or so mammalian species – excluding bats and marine mammals – are apportioned among mass classes. A corresponding analysis, but now restricted to the mammal species of Britain (and again excluding bats and marine mammals), is also shown in Fig. 4.8. Britain's

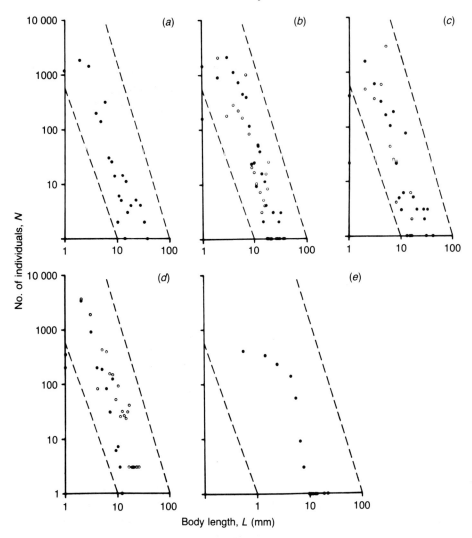

Fig. 4.7. The number of individual arthropods of different body
lengths, in a wide range of vegetation types, is related to body length
as predicted by combining metabolic considerations with aspects of
how organisms perceive their environment. In each case, the slope of
the relationship between log N and log L for the data plotted by
Lawton and co-workers lies between the predicted slopes
representing a 500-fold increase in N for a 10-fold decrease in L (lower
dashed line) and a 2000-fold (approx.) increase in N for a 10-fold
decrease in L (upper dashed line). (a) understory foliage in primary
forest in Costa Rica; (b) Osa secondary vegetation (solid dots) and
Kansas secondary vegetation (open dots); (c) Tobago primary riparian
vegetation (solid dots) and Icacos vegetation (open dots);
(d) understory foliage in cacao plantations in Dominica (solid dots) and
in Costa Rica (open dots); (e) birch trees at Skipwith Common, North
Yorkshire.

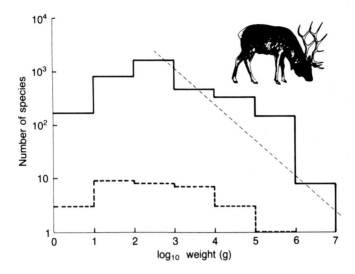

Fig. 4.8. The numbers of species, *S*, of all terrestrial mammals (solid histogram) and of British mammals (dashed histogram), excluding bats, are shown distributed according to mass catagories (mass expressed in grams). Note the doubly logarithmic scale. The thin dashed line illustrates the shape of the relation $S \sim M^{-2/3}$ (where *M* is the characteristic mass; this corresponds to $S \sim L^{-2}$, where *L* is the characteristic length).

mammals appear to obey the global pattern for number of mammalian species versus size, appropriately scaled down. Fig. 4.9 shows the corresponding species-size relation for butterflies in the Australian geographical realm, and in Britain. Again the British assembly appears to exhibit roughly the same species-size pattern as the Australian assembly, again scaled down by virtue of the smaller total number of species. Similar patterns have been documented for some other groups of insects, and for birds.

Fig. 4.10 gives a very crude estimate of the way in which the global totality of terrestrial animal species, from mites to elephants, are distributed according to characteristic length. As I emphasised when first publishing this work, the data presented in Fig. 4.10 are the result of a multitude of rough and uncertain estimates. The most serious problem is our current uncertainty, by a factor 10 or more, of the total number of species on the globe. I will return to this point in a moment.

Figs. 4.8–4.10 and other similar analyses represent rough assessments of the facts. Very few ideas have been advanced in explanation of these facts

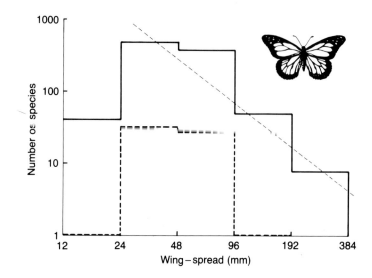

Fig. 4.9. The numbers of species, S, of butterflies in the Australian geographical realm (solid histogram) and in Britain (dashed histogram) are shown classified according to size (wingspread). The thin dashed line again corresponds to the relation $S \sim L^{-2}$, as in Fig. 4.8.

about species-size distributions. Interestingly, one of the few theoretical ventures into this arena is the only paper written jointly by Evelyn Hutchinson and Robert MacArthur, whom I regard as the two foremost ecologists of recent times. In this paper, they advance arguments for expecting the number of species, S, to be related to the characteristic length, L, of constituent individuals, by $S \sim L^{-2}$. This conjectured L^{-2}, or (given that $M \sim L^3$ as discussed earlier) $M^{-2/3}$, relationship is illustrated by the dashed lines in Figs. 4.8–4.10. The argument of Hutchinson and MacArthur is essentially that, for terrestrial organisms, the world is largely seen as two-dimensional, and therefore the possibility of finding new roles (and thence new species) scales as L^{-2} (that is, with area). Whatever the intrinsic merits of this suggestion, the work of Lawton and others (which we have just discussed) suggests several complications: for one thing, among insects at least, the environment is perceived as somewhere along the continuum from one-dimensional to two-dimensional; for another thing, the perceived environment scales with a fractal dimension, D, that is larger than unity (see above and Fig. 4.6(b)). Thus Hutchinson and MacArthur's argument should possibly translate to an expected relationship between S and L that is bounded by $S \sim L^{-D}$ and $S \sim L^{-2D}$, just as predicted for the relationship between number of *individuals* in a given size class and their size, as in Fig. 4.7. If D is roughly 1.5, this argument would lead to a predicted

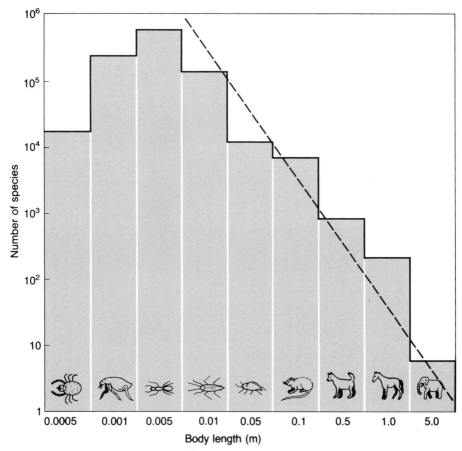

Fig. 4.10. A crude estimate of the distribution of all terrestrial animals, categorized according to characteristic length, L. Again, the dashed line indicates the relation $S \sim L^{-2}$, as in Fig. 4.8.

relationship between number of species, S, and length, L (the slope of the dashed line in Figs. 4.8–4.10), of between $S \sim L^{-1.5}$ and $S \sim L^{-3}$ rather than $S \sim L^{-2}$.

All this, however, is extremely conjectural. What we really need are both more empirical data about actual species-size relations in species assemblies in a range of geographical areas, and new ideas about possible mechanisms.

How many species are there on earth?

In studying patterns of species' relative abundance, or of numbers of individuals or of species versus physical size, one ultimate goal is to be able to understand things well enough to be able to estimate how many species there are likely to be in a given region. But, as things stand today, we do not even

Robert May

know – as a matter of fact – how many species there are on earth, much less
why it is this number, rather than a much smaller or a much larger one.

Until a few years ago, the uncertainties seemed somewhat less than they do
today, and the total number of species was widely believed to lie in the range
3–5 million. This estimate was obtained roughly as follows. For the species of
mammals, birds and other larger animals that are relatively well enumerated,
there are roughly two to three times more species in tropical regions than in
temperate ones. The total number of species actually named and catalogued
is just over 1 million. Most of these are insects. Indeed, to a good
approximation, all species are insects! But most insects that have actually
been named and taxonomically classified are from temperate zones. Thus, if
the ratio of numbers of tropical to temperate species is the same for insects as
for mammals and birds, we may expect there to be something like two or
three yet-unnamed species of tropical insects for every one named temperate
species. Hence the overall crude estimate of a total of roughly three times the
number of species currently classified, or around 3–5 million.

Terry Erwin's recent studies of the insect fauna in the canopy of tropical
trees has, however, resulted in a dramatic upward revision of this estimate, to
as high as maybe 30 million species.

Erwin used an insecticidal fog (generated by a machine hoisted into the
canopy) to 'knock down' the canopy insects. Because of their inaccessibility,
there have been few previous studies of the insect fauna in tropical canopies.
Erwin's studies suggest that most tropical arthropod species in fact live in the
tree tops, which is not so surprising as this is where sunshine, as well as most
green leaves, fruits, and flowers, are most abundant.

In studies of typical species of tropical trees, Erwin found more than
1100 species of canopy-dwelling beetles (including weevils), distributed
among the categories of herbivore, predator, fungivore and scavenger as
shown in Table 4.1. To use this information as a basis for estimating the total
number of insect species in the tropics, one needs to know what fraction of
the fauna are specific to the particular tree species or genus under study.

Table 4.1. *Erwin's estimates for the numbers of canopy beetles specific to a particular host-tree
species (Luehea seemannii), classified into trophic groups*

Trophic group	Number of species	Estimated fraction that are host-specific (%)	Estimated number of host-specific species
Herbivores	682	20	136
Predators	296	5	15
Fungivores	69	10	7
Scavengers	96	5	5
Total	1100 +		160 +

[78]

Unfortunately, as Erwin has emphasised, no data are available to provide a basis for estimating what this fraction is. As shown in Table 4.1, Erwin thus guessed that 20% of the herbivorous beetles were host-specific, 5% of the predators, 10% of the fungivores, and 5% of the scavengers. On this basis, he ended up with the estimate that, on average, about 160 canopy beetles are specific to a typical species of tree.

A variety of other assumptions are then needed to extrapolate from this estimate of 160 host-specific species of canopy beetles per tree species to the overall estimate of 30 million species in total. Slightly simplified, Erwin's argument runs as follows. First, beetles represent roughly 40% of all known arthropod species, leading to an estimate of 400 canopy arthropod species per tree species. Next, Erwin suggests the canopy fauna is at least twice as rich as the forest-floor fauna, and is composed mainly of different species; this increases the estimate to around 600 arthropod species that are specific to each species of tropical tree. Citing an estimate of 50 000 species of tropical trees, Erwin thus arrives at the possibility that there are 30 million tropical arthropod species in total.

It is easy to cavil at each step of this chain of argument. For instance, the fact that 40% of taxonomically classified arthropod species are beetles is of doubtful relevance if, in truth, essentially all arthropods are unclassified tropical canopy dwellers; what we need to know is what fraction of this canopy fauna consists of beetles. Furthermore, the overall estimate depends almost linearly on the necessarily arbitrary assumption that 20% of the herbivorous beetles are specific to a species of tree; changing this number to 10% would halve the global estimate to 15 million species. Nor is the estimate of 50 000 species of tropical trees at all secure. So little is known about many genera of tropical trees that it is often uncertain whether one is dealing with varieties or with distinct species; some tropical botanists prefer to use the more encompassing, and deliberately vague, notion of 'genus-group' for numbers of tropical trees. These reservations do not, however, detract from my admiration of Erwin's work, which has advanced us to the point where we can begin systematic investigation of each of the links in his chain of argument; Erwin's studies help to define a research agenda that will carry us toward a better estimate of how many species there are. In comparison, the earlier estimate of 3–5 million is pure hand-waving.

There is a different argument, altogether independent of any I have discussed so far, which suggests the total number of species exceeds 10 million. The plot of the number of species as a function of the physical size of the constituent individuals (on a log–log plot) in Fig. 4.10 gives an almost linear regression line, with a slope around −2, for species with characteristic linear dimensions above 1 cm or so. That is, for the relatively well-catalogued larger species, a ten-fold reduction in linear dimensions results in roughly a 100-fold increase in species; there are roughly 100 times more species with characteristic size 10 cm than with characteristic size 1 m. But as we move below 1 cm or so, the number of classified species falls further and further below the regression line that fits larger size categories. Were this regression extrapolated down to around 0.5 mm (0.0005 m), we would increase the species total from the known 1 million or so to more than 10 million. As we lack a fundamental understanding of the size-species relation itself, there is no reason to expect such a simple extrapolation to estimate the number of unclassified smaller species accurately. But it is interesting that the number is not inconsistent with Erwin's more biologically-based estimate.

Conclusion

One may imagine an extraterrestrial expedition landing on earth, and seeking to characterise our planet. Surely one of the first questions would be how many species of organisms there are. I find it truly extraordinary that our civilization – whose imagination reaches out to the boundaries and origin of the Universe, and in toward the structure of the nucleus and its constituent elementary particles – does not know, to within a factor of ten, how many species there are on earth.

Many readers of this book could, armed only with Avogadro's number, give a good estimate of how many atoms the book contains, thus quantifying a virtually unimaginable abstraction. The corresponding question of calculating, from first principles, how many species we expect to find in a given region or on the globe is undoubtedly a much harder question. But our ignorance of the simpler factual question of how many species we share our planet with stems not so much from inherent difficulty as from lack of effort, reflecting the vagaries of fashion and funding. We should commit more money and more effort to cataloguing our inheritance of biological diversity, and to providing an evolutionary understanding of the facts thus catalogued.

Further reading

Erwin, T.L. (1983). Beetles and other insects of tropical forest canopies at Manaus, Brazil, sampled by insecticidal fogging. In *Tropical Rain Forest: Ecology and Management* (eds. S.L. Sutton, T.C. Whitmore & A.C. Chadwick), Oxford: Blackwell, pp. 59–75.

Hutchinson, G.E. (1959). Homage to Santa Rosalia, or why are there so many kinds of animals? *Amer. Natur.,* **93,** 145–59.

Hutchinson, G.E. & MacArthur, R.H. (1959). A theoretical ecological model of size distributions among species of animals. *Amer. Natur.,* **93,** 117–25.

May, R.M. (1978). The dynamics and diversity of insect faunas. In *Diversity of Insect Faunas* (eds. L.A. Mound & N. Waloff), Oxford: Blackwell, pp. 188–204.

May, R.M. (1986). When two and two do not make four: nonlinear phenomena in ecology. *Proc. Roy. Soc.,* **B228,** 241–66.

May, R.M. & Seger, J. (1986). Ideas in ecology. *Amer. Sci.,* **74,** 256–67.

Morse, D.R., Lawton, J.H., Dodson, M.M. & Williamson, M.H. (1985). Fractal dimension of vegetation and the distribution of arthropod body lengths. *Nature,* **314,** 731–2.

Sugihara, G. (1980). Minimal community structure: an explanation of species abundance patterns. *Amer. Natur.,* **116,** 770–87.

[5]

Famine

Roger Whitehead

In recent years the subject of famine has been greatly sensationalised by the media. This chapter nevertheless attempts to approach the subject in an objective, academic manner. It stresses the importance of never under-estimating the nutritional complexities of the Third World and the *long-term* political, social, economic and technical developments that will be needed to avert the threat of famine which is always just over the horizon in the great majority of Third World countries.

The acute crisis

I shall not attempt to discuss modern famines within a historical perspective. It would have been intellectually attractive to ponder upon, say, the Biblical Famine, now ascribed the date of 1708 BC, in Genesis, Chapter 41, or perhaps on the West Bengal famine of AD 1770, or the more recent one of 1943. The potato famine of 1845 in Ireland would have provided equally good academic mileage, though the Africa famines in Nigeria of 1968 associated with the Biafran Civil War, or those of Ethiopia and the Sahelian region as a whole in 1972, 1978 and 1984–5 might have been of more immediate interest, but I am not going to discuss any of these famines.

My primary reason for not concentrating on such crisis situations is that it might reinforce the belief that the *single* catastrophic events which occurred on these dates, for example seven years of crop failure starting in 1708 BC in the Middle East or the peak of the drought in AD 1984 in Ethiopia, were the *primary* causes of these famines. Such a conclusion, although often reiterated, would be erroneous. One constantly reads that the three chief causes of famine are drought, war or pestilence. In reality these are often only terminal events tipping an already intolerable situation into one of complete disaster.

Coming to an oversimplified conclusion about the causes of severe undernutrition in an area is dangerous because a single-component, aetiological concept tends to breed only a single-component policy for the solution of that problem. Indeed in cases where drought or war or pestilence is deemed to be the only issue that matters, national and international planners have, not infrequently, used this as an excuse to spare themselves the responsibility of devising a solution. Usually it does start to rain again, as it has done in the Sahelian regions of Africa and in Ethiopia. Civil wars do ultimately drag towards some sort of victory or, more likely, stalemate. Biological pests mysteriously disappear again. Such natural restorations of the *status quo* can be greatly to the comfort of international leaders who may pride themselves in the belief that the art of good government, like good management, is to do nothing and most problems will correct themselves. On the surface this may be true. Who has any *political* cause to worry about Biafra now? The Ethiopians from Tigre, also, are leaving the refugee camps and returning to their tribal areas.

It is necessary, however, to look more deeply into this matter. What the crisis events have usually done has been to effect a negative shift in socio-economic status, in the distribution patterns of food availability or in susceptibility to infection and disease, etc. Because the vulnerable popu-lations are already on a health knife-edge the resultant deterioration is sufficiently rapid and severe to make dramatic photographic records for TV and the press, bringing publicity and temporary discomfort to both national leaders and remote observers in the wealthy countries of the world. When the acute crisis is over, the *status quo* knife-edge might be restored but it is unlikely that the basic underlying human problems will have been improved. Marasmus is still a major health problem in Ethiopia. Kwashiorkor continues to be extremely common in Biafra.

Status quo and the Third World

Health statistics

In most Third World countries, widespread sub-clinical malnutrition exists all the time such that even a short period of, say, civil disorder or drought can precipitate vulnerable members of the community into nutritional disaster. This fact is by no means as well appreciated as it should be, even by 'informed' international diplomats or civil servants. This is unfortunate because the international agencies do their best to provide accurate com-parative data on parameters such as infant mortality, proportions of underweight people within the community and food production per head of population (Table 5.1 and Fig. 5.1). For example, the infant mortality rate per 1000 live births in the 34 lowest income countries, including China, is 114 while in the UK it is only 11, a ten-fold difference. Likewise, at any one given moment in time, 2% of young children in the Third World are severely underweight (less than 70% of the weight they should be for their height) and a further 20% would be classified as mildly or moderately underweight (between 70 and 90% weight for height). Relating these percentages to actual

Table 5.1. *Statistics relevant to famine susceptibility, summarised from population reference bureau (1983), Washington*

	Population doubling time (yrs)	Infant mortality rate (per 1000)	Dietary energy supplies* (% of needs)	GNP in 1981 (US$ per capita)
AFRICA				
North Africa	23	109	110	783
West Africa	23	139	96	601
East Africa	23	111	88	305
Central Africa	26	121	95	483
Southern Africa	26	97	115	2349
ASIA				
Southwest Asia	26	99	113	3865
Central South Asia	30	124	91	251
Southeast Asia	33	85	103	663
East Asia	48	41	106	1396
THE AMERICAS				
North America	94**	11	187	12 405
Central America	26	58	115	1953
Caribbean	38	61	103	—
South America (tropical)	29	73	103	2065
South America (temperate)	47	39	122	2578
EUROPE				
Northern Europe	391	11	130	9935
Western Europe	436	10	134	12 704
Eastern Europe	125	21	136	4571
Southern Europe	141	19	139	5385
USSR	83	33	132	4701

* No allowances for inevitable food wastage has been made.
** Statistics influenced by immigration, not truly representative of birth rate.

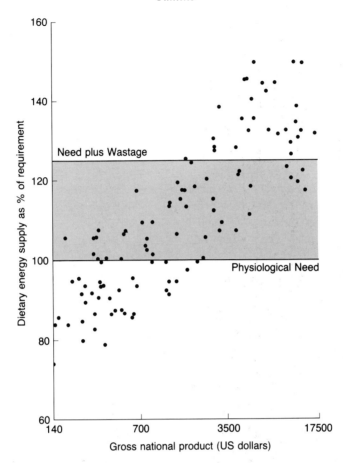

Fig. 5.1. The statistical relationship between per capita GNP of different countries in the world and the sufficiency of their dietary energy supplies. The lower limit of the shaded area represents the physiological requirement, the shading allows for realistic levels of food wastage. Based on *Data Sheet of the Population Reference Bureau (1983)*, Washington.

people, WHO has estimated that about 100 million children in the world currently have moderate or severe protein energy malnutrition. Quite clearly, these numbers are very much greater than the cases that get all the press publicity.

Physiological capacity for activity and work

Even these horrific statistics minimise the full impact of incipient malnutrition. Failure to be able to *function* adequately, to work effectively, is arguably more serious from both a national development point of view as well as for personal happiness and well-being. Clinical abnormalities and growth failure by no means tell the whole story, but this functional deficit, which Third World health workers have intuitively long known to be important, has unfortunately proved to be difficult to quantify with any degree of real accuracy. This is something which my colleagues are studying with the benefit of newly developed physiological technologies.

At this point it is helpful to consider a specific example of a typical Third World *status quo* 'at risk' situation. The MRC Dunn Nutrition Unit, where I work, is involved in nutritional health studies in the small West African country of The Gambia. While not a country which is normally associated with disasters, it is nevertheless a country where public health and nutrition are very far from ideal. It is one of the poorest countries in Africa and the official infant mortality rate, always a revealing nutritional parameter, is as high as 197 per 1000 live births. Yet to an increasing number of Europeans, The Gambia has become an exciting holiday centre run by healthy looking, handsome and friendly people. Superficially, such an impression is understandable, even for the very few tourists who do get away from the beach hotels to visit up-country. Fig. 5.2 shows a group of rural Gambian women clearing ground prior to the commencement of the agricultural season. It portrays an almost Arcadian vision of the Third World: it is the sort of thing a Victorian artist might have painted: certainly it is by no means a horror

Fig. 5.2. Gambian women clearing the ground prior to planting new food crops.

picture! Yet it is a prime example of a cursory snapshot giving a totally misleading impression.

The women in this photograph have all been part of our long-term research into pregnancy and lactation in rural Gambia. We have been particularly concerned to discover how such people manage to accommodate so well the dietary and other environmental circumstances which conventional scientific wisdom would declare impossible. On the negative side, we have also had to identify the point at which any postulated physiological and behavioural protective mechanisms might fail and essential function becomes impaired. In the Third World, scientific endeavour within the field of human biology cannot be focussed just on *optimal* health: defining the absolute minimum possible must frequently be the target for health planners.

Quantitative measures of body composition quickly reveal that Gambian women like those in Fig. 5.2 have remarkably small stores of body energy. A healthy, slim British woman would have about 13 kg of body fat and at the best time of the year, when food is not in particularly short supply, the fat content of the typical rural Gambian women too might occasionally be as high as this, but towards the end of each annual hungry season that fat content would typically fall to less than 8 kg.

Such a low body fat content is especially astonishing as all the women in Fig. 5.2 were either pregnant or lactating.

Nutritional stresses are particularly serious during the reproductive period of a woman's life. Fig. 5.3 shows how much maternal fat accumulation during pregnancy depends on the month when the Gambian mother conceives. In obstetric circles it is regarded as physiologically desirable to deposit an average of 4 kg of fat during a pregnancy, most of it during the second trimester of pregnancy. If the Gambian woman conceives in February and delivers in November, the second trimester and most of the third will fall within the hungry season and not only will she fail to lay down fat at all, on average she will lose 5 kg, a complete reversal of the natural physiological process.

Fig. 5.2 was taken at the beginning of July. Looking at it again, this photograph assumes a new significance. From now on, a particularly heavy agricultural work load over the next 4–5 months, coupled with a general shortage of food until the new harvest is in, will start to impose an increasingly severe strain on these women's energy physiology. The proportion of babies being born weighing less than 2.5 kg will as a consequence rise to as high as 30%, six times greater than one would expect in a healthy community. This smaller birthweight is associated with a higher perinatal and post-perinatal mortality and is a major contributory factor to the overall Gambian infant mortality value of 197 per 1000 live births already quoted. It is not just the women who lose weight during each annual hungry season; everyone, man, woman and child, who is dependent on their peasant economy, does so as well. The effect on babies under 2 years is especially marked owing to a reduced lactational performance also accompanying the dietary energy and nutrient stress of the hungry season (Fig. 5.4).

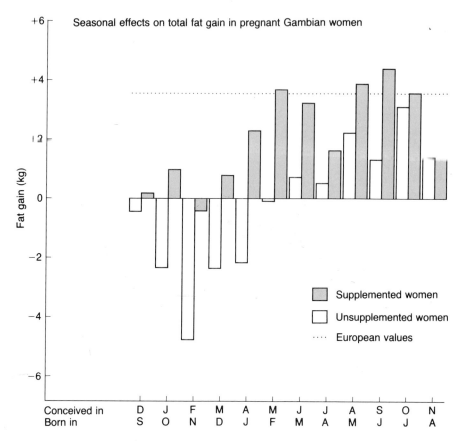

Fig. 5.3. Net fat accumulation (kg) during pregnancy among Gambian women depending on the month in which they conceived. The data show the responses of mothers receiving a dietary supplement as well as of those solely dependent on traditionally available foods.

Chronic undernutrition and acute crises

The three rural villages of Keneba, Manduar and Kantongkunda that we study in The Gambia are very much a microcosm of the sorts of problems facing the Third World as a whole. Around June to July *each* year in these villages the initial stages of potential disaster relentlessly appear. Only the appearance of early food crops starts to relieve this hunger, namely Findo (*Digitaria exilis*), a primitive cereal crop (also called hungry 'rice'). Only this and the short later period when maize becomes available avert a total disaster. It is easy to see why any factor interfering with crop-growing or harvesting at this critical time, such as a drought or civil war, would be so destructive and would rapidly precipitate an apparently stable community into a most severe state of famine. Such an event could very easily have happened as a result of the August coup d'état which was attempted a few years ago.

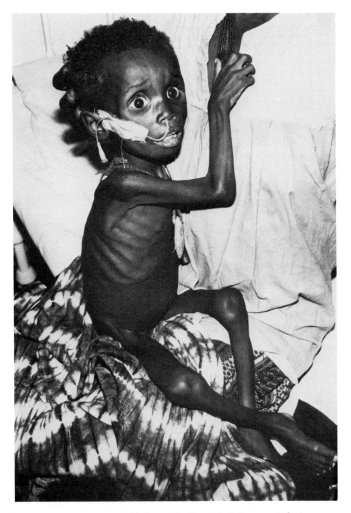

Fig. 5.4. A marasmic child from The Gambia being treated via naso-gastric tube feeding.

Knowledge like this undermines the value of focussing on specific acute famine events. Giving historical dates linked with particular crisis events might only divert attention away from the fundamental long-term nutritional issues affecting Third World countries.

Aid strategies

Recognising the importance of chronic sub-clinical undernutrition and malnutrition is more than an academic consideration. It has a direct practical bearing on the response of the richer countries of the world towards their less fortunate counterparts. It is an unfortunate fact that a significant proportion of the work of charitable organisations, particularly their fund raising efforts, has to depend on stop–go planning primarily modulated by the frequency of crises. Similarly, erratic variations in government aid programmes occur too. To some extent this is understandable. Charitable bodies and governments must be able to respond to emergency situations and provide appropriate aid as quickly as possible. It is equally clear that at such times both popular support for governmental action and the public profile of aid organisations will be high due to media coverage. People will consequently feel more inclined to give generously.

The danger, however, of an overdependence on 'crisis motivation' is gradually becoming recognised and we must welcome current attempts by the charities to re-educate their donors towards the necessity of supporting long-term development aid.

Working for change and improvement within the Third World can be a thankless task. Success, when it comes at all, comes only slowly and spasmodically. The only real hope is to influence future generations. The winds of rural conservatism are invariably adverse, as are the tides of governmental lethargy. The day-to-day activities of rural development workers in the Third World can be tedious and repetitious and one can only expect programmes to keep going if the 'enablers' in donor countries also accept a long-term commitment.

Grappling with crisis situations is inevitably much more glamorous and exciting. The aim of this review is to point out, however, that such activities, important though they may be at the time, will never solve the real underlying problems.

Appropriate long-term aid, national and local

It is not possible to offer a master-plan concept of what long-term aid strategy should always be as there can never be a single plan of action compatible with all situations.

Within any Third World country there are likely to be two philosophies towards food and health development. Frequently there is rivalry between these concepts but there is no real reason why this should be so.

Governmental aid programmes

Government-orientated aid programmes are likely to be on a very large scale and probably involve flamboyant projects such as dams, barrages or major irrigation schemes. Ultimately, however, the governmental approach is the one which has to succeed as it probably represents the only effective way to achieve changes of sufficient magnitude that will lead to lasting benefit.

The principal *claimed* governmental aim will probably be expressed as 'the generation of new wealth among the poorer sections of the community' and it is towards such an ethic that much of the international agency money goes too, as well as that of individual government aid programmes.

Any results are likely to be of delayed benefit only, however, and the sad experience has been that the socio-economic and health situation of the poor is rarely influenced to any demonstrable extent even within a medium-term survey period. The other problem with governmentally orientated programmes is that they can wax and wane according to short-term political expediencies, or they can end altogether via a coup d'état.

The charities

Aid from charitites is usually on a smaller scale, and relatively little of it contributes to a sustained improvement. What charities can achieve, however, is to bring interim help to where there is a crucial need. Additionally, and of equal importance, experience has shown that they can awaken and direct local community awareness towards health-related problems and thus sow the seeds of acceptance for programmes which are more encompassing. Organised from within the community, and less dependent on central governments, they tend to be more resilient in times of trouble. A notable example of this was in Uganda in 1971: while sophisticated programmes virtually collapsed, the small Nutrition Rehabilitation Unit set up by the Save The Children Fund (SCF) was still able to continue to provide its relatively low-key help and advice. As soon as there was some improvement in the general situaton, the Mwanamugimu Clinic, as it was called, could be revitalised much more easily than most other institutions.

Strategies for limited-scale rural development

What range of topics do localised programmes need to cover in order to fight malnutrition effectively and thus help to avert the spectre of famine? Practical, working demonstrations of ways of improving peasant food production and storage plus advice on the better use of resultant food products are going to be central themes, but it needs to be firmly borne in mind that food shortages are by no means the only factors which sow the seeds of widespread and severe undernutrition. Infection leads to at least as much malnutrition in the *status quo* situation as does a lack of food. Weight faltering and resultant severe growth failure can frequently be attributed to patterns of infection. Likewise, other clinical features such as oedema, the cardinal feature of kwashiorkor, which afflicted the children during the Biafran Civil War so markedly, has also been shown to be due as much to a succession of infections, such as respiratory tract infections and diarrhoeal diseases, as to primary protein deficiency.

The terms malnutrition and undernutrition are in many ways unfortunate in that they imply an overprecise causal diagnosis. The reality of the situation is that in the great majority of cases, syndromes like kwashiorkor and marasmus are multifactorial in origin and are products of a nutrition–infection complex.

The practical significance of this statement is that nutrition rehabilitation programmes, whether they are directed at the *status quo* situation or whether

Fig. 5.5. Traditional water well in The Gambia, West Africa. Such sources of water are prone to animal faecal contamination and are consequently an important cause of diarrhoeal disease.

they are dealing with a crisis situation, have to cover a much wider range of health issues than just ensuring food availability. All too often programmes are set up which focus too narrowly and which concentrate, say, on teaching mothers just how to prepare more nutritious weaning solids for their infants. The very best weaning foods will be totally useless if the nutrients fail to be utilised because of poor hygiene leading to infection (Figs. 5.5 and 5.6).

The value of food education in the Third World is often challenged because it tends not to lead to readily demonstrable improvements in nutritional status. The unresponding recipients of such advice, or their parents, are criticised, but it is frequently not they who are at fault but the food educators for not making their contribution part of a balanced health package. To return to the SCF Mwanamugimu Unit in Uganda: here the whole family of the undernourished child is approached and attempts are made to tackle a range of interconnecting issues such as the construction of more hygienic pit latrines, improved water supplies and advice is even given on ways of constructing more healthy living quarters cheaply. Naturally, food and nutritional improvements receive equal attention, as does child welfare, but it is the *balance* of the service the family receives which matters.

This breadth of approach is not just necessary in terms of *preventing* the

Fig. 5.6. Improved village water supply via a deep well and standpipe system in The Gambia.

development of sub-clinical undernutrition, it is just as vital for rehabilitating people after a severe famine. Getting food to people in conditions of severe nutritional hardship is of course crucial, but unless this is backed by the other types of help discussed above the long-term success rate will inevitably be low.

Food aid

In recent years a significant proportion of aid to the Third World has been provided in the form of food surpluses shipped from overproducing countries, to bridge some conceived nutritional shortage. This is done both during crises and on a more long-term basis.

Unfortuntely food aid has usually been organised in such a haphazard, *ad hoc*, and opportunistic manner that it has fallen, understandably but unnecessarily, into disrepute. While there clearly are situations where food aid is essential, the way such enterprises have been planned often leaves much to be desired. In a famine crisis the management of food aid can present logistic problems even more complicated than in the *status quo* situation, as the acute event will frequently have paralysed the whole infrastructure of a country.

It is no use sending surplus foodstuffs if it is impossible, under the circumstances prevailing, to utilise these for human feeding. For example, Europe and North America have well publicised wheat surpluses, but nobody eats wheat flour as such, and unless the emergency refugee camps are equipped with oven facilities for turning the flour into foods such as bread, much of this valuable commodity is likely to be lost.

Other food commodities, although they may be rich in the essential nutrients, may not be attractive to people if they are unused to that type of food. In Norway, fish powder and other by-products from the fishing industries can also be in surplus and have frequently been made available to the Third World as part of that country's food aid programme. However good the intention, however nutritious the product, no good will accrue if the population is reluctant to accept such a new foodstuff.

Dried skimmed milk is another contentious by-product of the European food industry, in this case from the butter manufacturing trade, and is produced in vast excess in Europe, including the United Kingdom. It too has been made available in large quantities to the Third World. There is no doubt about its nutritional potential. It is a good source of first-class protein, minerals and vitamins, but without modification it is useless as a food for direct consumption. Indeed, unless refortified with fatty sources of dietary energy similar to those originally removed when making the butter, feeding such a high-protein mixture as the sole source of food to undernourished people is ineffective, and in extreme circumstances, can be positively dangerous. For the nutritional therapy of young infants the nutritional potential of dried skimmed milk can be restored by the addition of vegetable oils and sugar. It is an unfortunate fact of life, however, that the donors of dried skimmed milk are rarely bothered about its effective utilisation: all too often it is merely dumped at the frontier and can become as hard as concrete if left in the sun and rain. Alternatively, it may be used to fatten-up rich men's pigs.

A further valuable food commodity which tends to end up being used inappropriately because of a lack of donor concern is the 'wheat–soy blend' mixture, a surplus product from the USA. This too does not readily fit into traditional patterns of household food preparation and, once more, all too frequently ends up being used inappropriately as a cereal mix for rich men's chickens.

The targeting of food aid

The misuse of dried skimmed milk and wheat-soy blend are particularly interesting examples because the Dunn Nutrition Unit has successfully used these two products in the formulation of a Gambian-produced nutritious biscuit specifically targeted at marginally nourished, pregnant Gambian women. Key components of this biscuit are locally produced groundnuts and groundnut oil, but together with the *rational* use of the two American and European aid foods, we were able to achieve a nutritious snack food which has proved remarkably effective in reversing the adverse effects of undernutrition on low birthweight in Gambian babies. As already described, during the hungry season, almost 30% of babies in our district of The Gambia are born too small by internationally accepted criteria, but the controlled use of our biscuit has reduced this to 5%, a value more typical of the richer areas of the world. As Fig. 5.3 shows, the biscuit also helped to minimise the abnormally low fat storage encountered in rural Gambian women.

A spin-off from these scientific studies has been the commercial development of a similar biscuit by the charity Oxfam for emergency famine aid use in Ethiopia and Sudan. Likewise, in previous decades, fundamental research into refortifying dried skimmed milk with vegetable oils and sugar for the intensive dietary therapy of babies severely malnourished with kwashiorkor, carried out by the British Medical Research Council, became the basis for UNICEF's worldwide hospital treatment schedule.

The essential principle is that food aid should be targeted towards people who truly are in need and should be delivered in a form capable of being utilized under the cultural, environmental *and* clinical circumstances which currently prevail within that particular country.

Food aid and the inhibition of local initiatives

The use of food aid on a long-term basis poses more complex, economic issues than if it is limited just to emergencies. Here the crucial issue is the balance between the *benefits* the external food can achieve, in terms of improving general health and well-being, and the *inhibitory effect* it might have on initiatives to diversify and improve local food production. This has been a subject of much political rhetoric partly because it has not been recognised that there can be no single strategy applicable to all situations at all these times.

Requests for extended food aid must be the responsibility of the nation in need and should be made only after all the relevant issues have been thoroughly explored. Quite clearly, the primary consideration must not be just to find a convenient dumping ground for the excesses of the affluent world. It is equally important, however, that political dogmas in the Third World about 'dependency' do not deny needy people food if this can be made available as part of a constructive programme of development.

A widely publicised attempt along these lines is the type of scheme which links developmental programmes with food aid by paying people for their labour with the aid food. An attractive example of this in The Gambia was run by the Action Aid organisation which encouraged one of the villages we study to build a system of simple earthen banks near their rice fields to improve water holding and hence rice production once the rains arrive. With rain being confined to a single period of the year, and rice production only being possible for this short time, such a scheme had obvious attractions. The currency of payment during construction was also in the form of rice, imported by the aid charity, a welcome commodity in the middle of the dry season!

Another and more recent variant on this theme is for aid food to be *sold* by government agents on the open market within the recipient country and for the proceeds to be used in funding relevant development projects. The theoretical advantage of such a venture is that the aid food becomes an intrinsic part of the commercial economy of the country rather than a charitable appendage. The financing of subsequent development programmes with local currency is also more under direct governmental control than is normally the case with externally financed schemes.

The Third Worlds

So far this chapter has considered the Third World as though it were a single entity. It is, of course, nothing of the sort. In some of the countries which still tend to be thus classified, there truly has been development. Parts of Asia in particular are definitely moving ahead. Countries such as Korea, Malaysia, Hong Kong and Singapore now occupy a position well up the World Bank's league table of per capita income. Importantly in these countries, malnutrition is now found just in pockets of poverty rather than as a generalised public health problem as was the situation in the past. Even in countries like India, which in 1983 had a *per capita* GNP of only 253 US dollars compared with the 1817 dollars of Malaysia and the 5220 dollars of Singapore, the 'green' agricultural revolution would appear to be resulting in at least some improvement in general food production.

The sub-continent of India

Community health workers and nutritional scientists in India argue, however, as to whether or not the green revolution really is having any beneficial effect on the poorer sectors of the community. The Indian National Nutrition Monitoring Bureau has produced data which did indicate some decline in the prevalence of *severe* malnutrition after around 1976 – a reduction in the proportion of children weighing less than 60% of the weight standard for age; but seemingly this has been without a corresponding shift in the growth status of the population as a whole. This has been interpreted as indicating a reduction in the tip of the iceberg of community malnutrition, probably mediated via nutrition intervention programmes and relief schemes aimed specifically at severely malnourished children, rather than a beneficial shift in the nutritional plight of the Indian poor as a whole.

Politicians the world over make optimistic claims for the effectiveness of their national programmes, especially near election times, that do not measure up to critical scientific evaluation. When scientists point out such discrepancies they are inevitably unpopular and are accused of being negative

Fig. 5.7. A markedly underweight man from **Bangladesh** eking an existence from the sale of meagre stocks of vegetables. Note, however, the cigarette.

thinkers. There is more than an element of despair in a recent pronouncement by the world famous Indian nutritionist, Dr C. Gopalan, who says:

> Endless (diversionary) debates and the tiresome search for alibis and fig leaves have tended to obscure the fact that by whatever reckoning the problem of undernutrition in our Indian children and women is massive and continues to progressively erode the quality of our human resources and despite the fact that our constitution specifically lays down nutritional upliftment of the people as a national objective, six Five-Year-Plans have failed to make a significant dent on the problem.

For countries for which data on gross national product per head of population are available, the poorest Asian countries in 1983 were Bangladesh

Fig. 5.8. A group of children photographed in the rain in the slums of
Dacca, Bangladesh. The young child in the centre is markedly
marasmic and he is also suffering from severe vitamin D deficiency.
The older boy on the right exhibits angular stomatitis resulting from
riboflavin deficiency.

(144 dollars) (Figs. 5.7 and 5.8), Nepal (156 dollars), Burma (183 dollars) and
India (253 dollars). The regional consequences of this poverty in terms of
infant mortality rate are substantial. The values in the countries just cited
range from 99 to 149 per 1000 live births in contrast to 11.7 quoted in 1983 for
Singapore, 9.1 for Taiwan and 7.1 for Japan. These latter data are better even
than the corresponding UK value.

The relative poverty and poor health statistics of the Indian sub-continent
as a whole stand out even when compared with other major areas of the Third
World. In Latin America as a whole, for example, infant mortality rate is
'only' 65 per 1000 live births, passing 100 only in Haiti (113) and Bolivia (130).
The only geographical region which offers any serious rivalry in the poverty
and ill-health stakes is Africa.

[98]

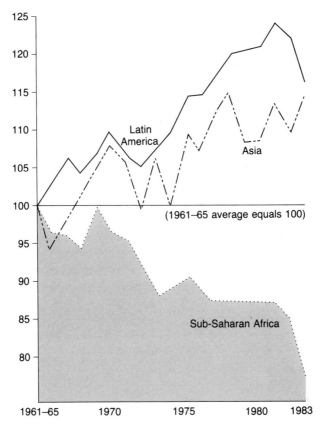

Fig. 5.9. *Per capita* food production trends in Asia, Latin America and Africa relative to levels in the early 1960s.

Africa

In what is conventionally known as 'black Africa south of the Sahara and north of the Zambezi', national values for infant mortality rates are always above 100 per 1000 live births, especially in West Africa where rates of 206 and 201, for example, were recorded in 1983 in Sierra Leone and Upper Volta. This was even before the most recent Ethiopian and Sahelian drought: indeed in this chapter I have limited my statistics to 1983 in order to avoid this complication.

There are very special reasons for nutritional concern in Africa. In Asia, even in the sub-continent of India, there is no real sign of the basic food *supply* actually getting worse. Indeed Fig. 5.9, redrawn from data supplied by the US Department of Agriculture, indicates that since 1960, *per capita* food production in Asia has, in general, shown an overall upward trend. This improvement has not been on quite the scale calculated for Latin America but, quite clearly, is infinitely better than the average for the 39 African countries south of the Sahara (excluding the Republic of South Africa and Namibia) where the trend was very definitely downwards. For these 39 countries as a whole *per capita* food availability by 1983 was down to 80% of what it had been at independence in the early 1960s.

In much of Africa, cereals are the main dietary staple and an analysis by the United Nations Food and Agriculture Organisation suggests that over a similar time period *per capita* cereal production in black Africa had fallen by almost 60%. Maurice Strong, the former Head of the United Nations Environment Programme, has described the situation we are witnessing in Africa as the 'greatest eco-catastrophy in recorded history – the greatest example of what happens when the balance between environment and development breaks down'.

Africa is the only continent failing to keep food production ahead of population growth.

Food production versus population growth in Africa

Inevitably, when one discusses nationwide food shortages and balancing the equation between production and need, the issue of population control becomes one of the key foci of attention. At a growth rate of 3% per annum, the United Nations have estimated that black Africa is likely to increase in population from a current 384 million to 645 million by the year 2000 and 1.27 billion (10^9) by 2025. Such a population growth rate is quite out of step with current rates of increase in indigenous food production as well as exceeding any conceivable potential for improvement within existing agricultural strategies. Left to the processes of natural selection, the population explosion will ultimately be limited by widespread malnutrition associated with even higher levels of infant mortality than those found at the present time.

Population control and family planning

An alternative to increasing food production is to try to reduce birthrate. Population control and family planning are not such an immediately attractive proposition for most Africans, however, either at a governmental level or within individual families. Overcrowding is not so obviously overwhelming as it is in parts of Asia. Bare statistics do not tell the complete story adequately, but in Africa as a whole there are only 18 persons per square kilometre while the equivalent value for Asia is 80 persons. Even in relatively fertile parts of Africa such as Uganda and The Gambia, for the casual observer as well as the rural Gambian, there is apparent scope both for a substantially increased population *and* greatly improved food production.

One should not dismiss too quickly the reasons why rural Africans are so conservative. They are, for example, totally dependent on their families for support when they grow old: reducing their familial size would be the desperate equivalent of our cashing in a life insurance policy before it had

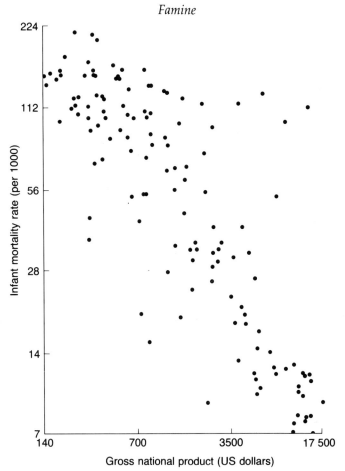

Fig. 5.10. Relationship between infant mortality (per 1000 live births) in different countries of the world and per capita GNP. Data source as in Fig. 5.1.

matured! At a more insidious level there is always, not too far below the surface, the suspicion that population control programmes are really a ploy to limit the impact of the black nations on the rest of the world.

Child welfare schemes and family planning

Population control in a free society can only be mediated through voluntary family planning and in Africa this will only be accepted if child health care systems can be sufficiently funded to ensure the survival of children who are already born. Fig. 5.10 shows the relationship between national *per capita* wealth and infant mortality. Extra children are an insurance against future losses by death.

Effective stabilisation of population size is a natural concomitant of a growing sense of general well-being: it is difficult to impose family planning in the absence of this confidence for the future. One frequently hears

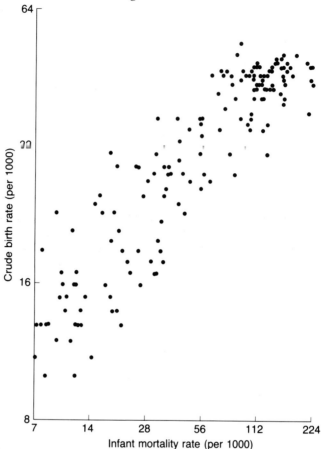

Fig. 5.11. Relationship between national infant mortality rates (per 1000 live births) and corresponding annual birth rates (per 1000 of the population). Data source as in Fig. 5.1.

comments to the effect that putting money into child health care schemes would merely make a bad situation even worse by permitting more children to live rather than to die. Ignoring the inhumanity of such a philosophy, child health care schemes, arguably, are the basis for the only population control system that has any practical chance of succeeding. Understandably there might be a population 'overshoot' when child mortality rates initially fall below birth rates but surely it would be permissible to extrapolate between what has happened during the last century in Europe and the United States into future Africa? Fig. 5.11 shows the close relationship between birth rate and infant mortality: the lower the mortality the lower birth rate. For rich confident western Europe the calculated population doubling time is now 436 years, in the UK it is 533 and in Sweden an incredible 3465 years but in Asia as a whole it is only 36 years and in Africa just 23 years.

Facile cliché 'solutions' are inadequate. To borrow a clinical euphemism, it is necessary to treat the disease not the symptoms. Economic stability and family well-being are essential for both sides of the food production–population equation.

AIDS

Another current fallacy is that the AIDS epidemic will solve the African food problem because of the very high incidence of the disease in some parts of that continent. In a lecture to the British Nutrition Foundation, Sir Kenneth Blaxter refuted this suggestion and claimed that a million AIDS deaths would make very little impact on population dynamics as a whole. This is quite true, but a further important fact which was not taken into account was the age distribution of the disease. It affects primarily the sexually active sector of the community which essentially means the young 'bread-winners'. Rather than help the food equation, the rampant progress of AIDS will almost certainly exacerbate food production problems much in the same way that war, by killing off able young men in 1914–18 and 1939–45, has greatly impeded national development in eastern Europe. With AIDS, however, women will not be able to cover for the lost men as they have had to do in Russia. They too will be dying prematurely at the peak of their economic potential.

Food production

The debates about the relative merits of high-technology farming methods versus the so-called 'appropriate technology' approaches lie outside the scope of this chapter. Whichever approach is taken it is going to cost money and will have to involve both a concerted national as well as international effort. When the Israelis started to colonise Palestine, they were able to make dramatic agricultural strides because of massive financial help from the USA, and if the African peoples eventually show an equivalent political will, they too will require similarly generous financial backing.

Predictably they will have to resist external pressures, as well as internal desires for self-aggrandisement, and becoming involved in a plethora of prestigious show-projects. In some places large dams will be necessary, but their precise role in the agricultural strategy of a country will need to be well thought out and not embarked upon just because external donors are prepared to lend money for this and seemingly no other agricultural enterprise.

Food and agriculture research

Although food research, particularly plant and animal breeding, must form a key part of any internationally promoted long-term emergency package for African agriculture, this needs to be directed at the specific crop needs and crop problems of Africa and not just be a by-product of European or American applied research, transplanted inappropriately into a totally different economic and cultural climate.

To be more specific, African countries have seen a number of attempts by aid donors to introduce into their agricultural economies foreign food crops such as the soy bean. This is a profitable crop in the USA, but in all too many of the African countries where this has been done, there has not been any real need for such a food commodity. Existing food crops, like the groundnut, frequently filled the same agricultural role perfectly well and research and development would have been better focussed on this and other indigenous crops.

Likewise, vast amounts of money have been spent on plant breeding schemes to produce genetic cereal variants such as high-lysine maize. This was done for seemingly sound nutritional reasons but there are more simple ways of ensuring a balanced amino acid mixture in the diet. Disappointingly, there has been very little intensive effort to improve the nutritional quality of *low*-protein staples like cassava, yams, and plantains as well as traditional cereals such as sorghum and millet. Intensive research into new drought-resistant types of foods is also critically required.

The way in which education in basic agricultural practices is provided also deserves consideration. Perhaps European farm labourers might provide a more effective interface with the African peasant farmer than the well-educated agriculturalists who presently attempt the task over an unbridgeable intellectual divide.

Finance

Financing any African blueprint for survival must involve more than finding *new* money. It will need a totally revised attitude towards existing national as well as international monetary priorities.

Arms expenditure

The way that civil wars have severely strained an already difficult situation in Africa and precipitated famine has already been mentioned, but military activities play a more insidious long-term role by diverting capital away from the true action necessary in the war for national survival. In all countries with armies, there is debate about the relative proportions of the national resources which should be spent on 'defence' on the one hand and on health, education and development on the other. The overwhelming bias needed for Africa is quite clear, but one almost invariably finds the most advanced and expensive technologies sequestrated in the military forces. Spending money on products of non-developmental value obviously needs to be drastically curtailed but this will require financial discipline both on the part of African leaders and on that of their commercial suppliers in Europe and America. The main arms suppliers to sub-Saharan Africa between 1970 and 1980 were the

Soviet Union, who accounted for 50%, followed by France at around 15%, with most of the remainder coming from a range of NATO countries including the UK.

Interest debt

There are additional drains on African economies which are making it impossible to reverse the poverty spiral. Twenty-two of the worlds 36 poorest countries are in Africa. For the 1970s, available statistics indicate that rather than showing a rise in GNP the economies of 15 African countries actually fell. Even more serious, most have allowed themselves, or been encouraged by foreign governments both east and west, as well as by the international agencies and private banks, to get into a hopeless credit debt. Paying back the interest alone on these debts is frequently more than equal to income from overseas aid *plus* local industry.

It is not easy to come up with a convincing long-term plan to reverse the deteriorating situation in Africa, but sooner or later the World is going to have to get its 'act' together if a massive crisis in Africa is to be avoided. It will require international cooperation both between the east and west as well as across the emotional barriers of 'north–south' divide. The long-term solution *has* to be an economic one. Some way of overcoming the stranglehold of interest repayments will have to be devised or there will be no hope for significant indigenous investment in development programmes.

Most crucial of all, there will have to be a totally different attitude of mind *within* Africa itself. The old rivalries of tribe, black versus black, black versus white, hang-ups over 'colonialism', communism versus capitalism, christianity versus mohammedism, all will have to be buried if Africa is to be placed back onto a road of improving prosperity.

Conclusion

This chapter has been about famine but, as stated at the outset, it has not been about specific acute famine situations. My concern has been more towards analysing the chronic environmental, cultural and economic factors that make countries prone to crisis after crisis and famine after famine rather than to the short-term acute events which, erroneously, are often assumed to be the major cause of a particular famine. Economic development and helping people to create a more stable, prosperous basis for life and living must be the target if we are to halt the advance of famine in Africa and other impoverished parts of the world.

Further reading

Alleyne, G.A.O., Hay, R.W., Picou, D.I., Stanfield, J.P. & Whitehead, R.G. (1977). *Protein Energy Malnutrition*. London: Edward Arnold.

Amann, V.F., Belshaw, D.G.R., Stanfield, J.P. (1972). *Nutrition and Food in an African Economy*, 2 vols. Kampala: Makerere University Press.

Biswas, M. & Pinstrup-Andersen, P. (1985). *Nutrition and Development*. Oxford: Oxford University Press.

Harper, A.E. & Davis, G.K. (1981). *Nutrition in Health and Disease and International Development*, Vol. 77. New York: Alan R. Liss Inc.

Hoorweg, J. & McDowell, I. (1979). *Evaluation of Nutrition Education in Africa*. The Hague: Mouton Publishers.

Manning, D.H. (1976). *Disaster Technology*. London: Pergamon Press.

Passmore, R. & Eastwood, M.A. (1986). *Human Nutrition and Dietetics*. Edinburgh: Churchill Livingstone, Chs. 57–9.

Royal Society of London (1977). *Technologies for Rural Health*. London: Royal Society.

Schofield, S. (1979). *Development and the Problems of Village Nutrition*. London: Croom Helm.

Schurch, B. (1983). *Evaluation of Nutrition Education in Third World Communities*, ed. B. Schurch. Bern: Hans Huber Publishers.

UNICEF/WFP (1985). *Food Aid and the Well-Being of Children in the Developing World*. New York: United Nations Children's Fund.

[6]

Exhaustible resources

Partha Dasgupta

Introduction

All human activity is ultimately based on resources found in nature. Whether it is consumption, production, or exchange, the commodities which are involved can always be traced to constituents provided by nature.

A tractor requires for its manufacture iron and steel, rubber, plastics, non-ferrous metals, labour of various skills, factory-machinery, water, and so on. Of these, to take an example, the steel requires for its production iron-ore, coal, furnaces, water, labour, and so forth. The furnaces in turn require for their manufacture, among other things, iron ore, brickworks, and labour. One can thus break down any produced good into the inputs involved in its manufacture and one can, if one has sufficient patience, trace them ultimately to a combination of labour and natural resources. Of course, labour too is produced and sustained by natural resources. So ultimately all commodities and services can be traced to natural resources.

My purpose in reminding you of the morphology of produced goods and services is not to prepare a chapter on the resource-theory of value. Like Marx's labour theory, such a construct quickly runs into analytical difficulties. My purpose, rather, is to set a materialistic tone so that one may in an unhampered way view natural resources in the light of their *use* to us in running our lives. I am not suggesting that this is the only defendable perspective; I am merely indicating that this is the attitude I shall strike in this chapter. I need to declare my position at the outset, because if you scratch a resource economist you are likely to find a nature-worshipper trying to get out. I am going to suppress this urge and concentrate instead on the *instrumental* value of natural resources.

It is usual, and convenient, to divide natural resources into two classes; *exhaustible* (such as coal, oil and natural gas) and *renewable* (such as fisheries, aquifers, forests and woodlands, and soil quality). On the face of it, the classification is misleading, because even renewable natural resources are in danger of exhaustion if they are harvested at too fast a rate (see chapters by Goudie, Myers, and Wells, this volume). But the idea behind the distinction is clear enough. Resources such as fossil fuels enjoy growth rates which operate only over geological time. Thus, their stock cannot increase. It can only decrease, or if none is mined for a while, stay constant. Improvement in technology – for example, learning to drill offshore – can increase the *usable* stock; this is a different matter. Then again, discoveries of new deposits increase the *known* available stock. That, too, is a different matter. Now, what I have said about fossil fuels is true also of recyclable material such as metals. The redoubtable Second Law of Thermodynamics ensures that we will never recover an entire ton of secondary copper from a ton of primary copper in use, or an entire ton of tertiary copper from a ton of secondary copper in use. There is leakage at every round, and a simple formula, based on compound-decay, tells us how much copper can be used from the initial stock once you know the recovery, or recycling, ratio. Since the formula is easy to derive I am presenting it in Table 6.1. In Table 6.1, I assume that 100 tons of an ore are initially put to use, and that recycling occurs at the beginning of every period. Now, the recycling ratio is in fact subject to choice, the higher the ratio sought the greater is the cost of recycling. The key point though is that the ratio can never be unity. For simplicity of exposition I avoid all the complications that such choice introduces and assume merely that the recycling ratio is a fixed fraction, 0.8. Thus, in the second period only 80 tons remains in use, in the third period $0.8 \times 80 = 64$ tons, and so on. Adding all these figures yields an effective initial stock of 500 tons, five times greater than the nominal stock of 100 tons. But even 500 is finite. It is as though we are given 100 tons to use in the first period, 80 tons in the second period, 64 tons in the third, and so on, going down to vanishingly small figures in the long run. So metals and ores are also exhaustible resources.

In their chapters, Andrew Goudie, Norman Myers, Robert May and Gordon Wells have written on renewable natural resources, such as soil, forests and animal species. I shall therefore confine my discussion to exhaustible natural resources in the sense in which I have defined them.

History

Fear of the impending exhaustion of natural resources would seem to be a periodic state of mind – and not always prompted by good reasons. The worldwide concern over the availability of fossil fuels in late 1973 was in great part occasioned by the four-fold price increase in Gulf oil. But this increase ought not to have been seen as a signal for increased resource scarcity.

Table 6.1. *Expansion of resource base through recycling*

Assume that there are 100 tons of the ore, which are recycled each 'year' at a recycling ratio of 80%. Then total effective stock is:

$$100 + 0.8 \times 100 + 0.8 \times 0.8 \times 100 + 0.8 \times 0.8 \times 0.8 \times 100 + \ldots$$
$$= 100 \{1 + 0.8 + (0.8)^2 + (0.8)^3 + \ldots\}$$
$$= 100/(1 - 0.8) = 100/0.2 = 500 \text{ tons}$$

Worldwide reserves had not suddenly shrunk. Nor had demand jumped all at once without notice. The rise in price was, rather, the exercise by OPEC as a cartel of its market power. The consequences to the world economy have been enormous, as Michael Bruno of the Bank of Israel and Jeffrey Sachs of Harvard University, have recently documented (see Bruno and Sachs, 1985). But it had nothing to do with resource exhaustion. You can see this vividly in the fact that, with the weakening in cohesion among OPEC members, energy scarcity has fallen from the agenda of public anxiety.

This is not to suggest that societies have not faced resource scarcities in the past, nor that they have not done something about it. Man has been exhausting the earth's resources ever since he began walking around and making fires (see Goudie, this volume) and he will be doing so for a long while yet. Indeed, if one is in that sort of mood one can look at history as a relentless pursuit of new, substitute, resource bases for existing, vanishing ones. These substitute resources are sometimes present in ready-made form. Sometimes they are not. In either case it is a matter of devising ways of *using* them, and by this I mean *economically* feasible ways of using them. This is the hard part. It involves research and development: R&D. The point I am trying to highlight is the banal fact that the basic ingredients of all present and future commodities and services are present today. It is only when we learn to make use of them, or when we start expecting to make use of them, that they attain some economic value. Then we take an interest in them. To put it in the words of E.W. Zimmerman: 'Resources are not, they become.'

To cite a concrete episode, a number of historians have seen the Industrial Revolution as a response to a concern with the supply of raw materials, especially sources of energy. Preindustrial societies were dependent upon wind, water and animal power. Above all, they were dependent upon *wood*, not only as a building material, as an industrial raw material and as a source of chemical inputs, but also as a source of *fuel*. The need for finding substitutes for wood-fuel was being signalled in England as early as the Elizabethan period. The signal was a secular increase in fuel prices, which, as it happens, rose during this period three times as fast as prices generally. The Industrial Revolution in Britain involved a substitution of cheap coal for wood as a source of fuel and power and an abundant source of iron for vanishing timber resources. The process was slow, because this substitution was fraught with technical difficulties, but it happened.

An extreme version of this viewpoint was expressed by the economic historian Brinley Thomas some years ago:

> In the second half [of the eighteenth century] Britain ex-
> perienced an energy crisis . . . there was a growing general
> shortage of timber and especially charcoal, whereas coal was
> relatively plentiful . . . [the problem] could only be resolved
> by switching the energy base from wood to coal . . . but this
> could not be done until a fundamental innovation had been
> made . . At the beginning of the 1780s Britain was in trouble,
> a special kind of trouble which no other country faced at that
> time . . . she was in a severe energy crisis, dangerously

dependent on wood fuel . . . In the previous twenty five years as many as thirteen inventors had attempted to overcome the technical difficulties. It was not until 1784 that the problem of refining pig iron with coal or coke was finally solved by Henry Cort's process . . . the Industrial Revolution occurred in Britain at the end of the eighteenth century not because Britain was 'well-endowed' in various respects, but because she was 'ill-endowed' in one fundamental respect – she was running out of energy and had to do something about it. France had no such problem.

(Thomas, 1980)

The point of view is single-minded, but the thesis is at the very least suggestive.

In the United States, where resource endowments were quite different, the temporal pattern of resource-use was also different. In the early part of the nineteenth century, wood in America constituted an abundant and cheap fuel. Consequently, its use continued long after England had ceased to rely on it and had adopted coal. In 1850, mineral fuels supplied less than 10% of all fuel-based energy in the United States, whereas wood supplied 90%. By 1915 wood had declined to less than 10%, whereas coal accounted for 75%. The story since the end of the First World War has largely been one of a move away from coal to oil and natural gas as sources of energy (see Chapter 14 of Rosenberg, 1976, for an elaboration of this).

Resource substitution as the key process

This example of alternative energy sources and the reasons underlying a move from one source to another with the passage of time is indicative of an elementary but important truth. Demand for minerals is for the most part derived demand; that is, it is derived from our demand for final goods and services. Now, it is not the minerals *per se* which are necessary for the production of the final goods and services; it is certain key *properties*, or *characteristics* that are required, and it is these we seek in minerals. The properties I have in mind are familiar; for example, tensile strength, conductivity – or lack of it – durability, porosity, texture, specific gravity – high or low – colour, magnetism, and so forth. To take an example discussed by Anthony Scott (1962) in a most interesting article on resource substitution in the past, no one cares for the metal 'tin' as such. One *does* care for certain of its properties. It makes copper harder and iron resistant to corrosion. It is also durable, light and malleable, and this makes it a convenient constituent of containers for food. Each of these functions can now be performed by other means. Tin's hardening of copper can be achieved by other metals. Glass jars have replaced tin cans as food-containers in many circumstances. The basic constituent for glass is sand, available in vast quantities, and glass requires for its manufacture less expenditure of energy than does the extraction of tin, when the latter is sought in low-grade deposits containing minute quantities of the stuff. This is precisely why glass jars are replacing tin cans. Lead and mercury are being replaced in a similar way. And one could go on.

What are the innovative mechanisms, the characteristics of technological change, which bring forth such resource substitution as I have been discussing? They would appear to be seven in number, often overlapping, and often simultaneously occurring.

There is, firstly, the sort of innovation which enables a given resource to be used for a given purpose, that is, to use Zimmerman's aphorism once again, the kind of innovation which enables a resource 'to become'. The development of techniques for using coal in the refining of pig iron is a classic example of this. Research and development aimed at the control of nuclear fusion is designed to bring forth this sort of innovation. There is, secondly, the development of new materials, such as synthetic fibres. Thirdly, there are technological developments which increase the productivity of extraction processes; or in other words, make extraction of certain minerals cheaper. In the early part of this century the manufacture of large earth-moving equipment made the strip-mining of very low-grade ore deposits possible. Fourthly, there are scientific and technological discoveries which make exploratory activities cheaper. Developments in aerial photography and seismology have been crucial here. Fifthly, there are technological developments which increase efficiency in the use of resources. To give an example, during the period 1900 to the 1960s, the quantity of coal required to generate a kilowatt-hour of electricity fell from nearly seven pounds to less than one pound, a more than seven-fold reduction. Sixthly, there are developments of techniques which enable one to exploit low-grade, but abundantly available, deposits. For example, the discovery of the technique of froth floatation allowed low-grade sulphide ores to be concentrated in an economical manner. Seventhly, there are constant developments in techniques of recycling, increasing the recycling ratio at a lower cost, thereby raising the effective stock in the manner I indicated earlier, in Table 6.1.

There is in fact an eighth mechanism, much more prosaic, and unconnected with any *new* technological development. It consists simply of the substitution of lower-grade resources for vanishing high-grade deposits, where the move involves a more costly extraction process, one which has been available all along. In early years, when the high-grade deposits are amply available there is no need to exploit lower-grade ores. But, being exhaustible, the superior deposits eventually get depleted. This is then the time to move on to inferior deposits, for which extraction and refining costs are higher. Examples of this abound.

(I should add parenthetically that economic theorists have also explored a ninth possible mechanism, the substition of fixed, produced capital, e.g. factories, for vanishing resource use. However, substitution possibilities along this route are severely limited, so I shall not consider them further here.)

All this looks like pure taxonomy, and taxonomy induces boredom.

Fortunately, there is a single organising idea behind all these substitution mechanisms, and this enables one to study the economic process of resource exhaustion and substitution within a single analytical framework.

It can be shown that all the first seven mechanisms can be understood in terms of the eighth. There is thus a single analytical framework which encompasses all the eight mechanisms I have listed above. In particular, the eighth mechanism is a special case of each of the first seven. This is comforting, because even at first sight the eighth mechanism is the simplest to understand: it alone does not involve the production and use of new knowledge.

As it happens, the economic analysis of the eighth mechanism had been presented, and presented to all intents and purposes in a complete form, in a 1931 article by a great statistician–economist, the late Harold Hotelling. His classic article, which reads as though it was written last week, went pretty much unread, presumably because the timing was wrong – the beginning of the Great Depression. In any event, no progress in the analytical economics of exhaustible resources was made for over 40 years subsequent to the publication of Hotelling's classic.

Now, this reductionism – the reduction of the first seven economic mechanisms to the eighth – may appear to be a sleight of hand. New knowledge as required by the first seven economic mechanisms must by definition be unpredictable in some sense, and in extreme cases one may not even have a language to discuss it prior to its acquisition. So then how can one analyse and plan for resource exhaustion and substitution on the basis of the first seven-fold taxonomy? The answer is that in discussing the means of resource substitution we are not required to concern ourselves with the *exact* nature of future discoveries. If we needed to we would be caught in the sort of dilemma the Greeks found irresistible: to know now something we will only know in the future. So of course we do not try to do that. What we try to do instead is to identify the economic characteristics of *possible* discoveries. Thus, for example, we may not now possess the technology for mining deep-sea nodules from the middle of the Pacific, but technologists may have good reasons to be sanguine that if so much research attention is given to the problem, within 20 years' time there is a good chance that economically viable techniques will have been developed. The key term here is 'economically viable' and one can easily see that the characteristics of certain operations are the relevant things, to wit: finding cheap ways of operating equipment several miles below the ocean's surface over vast tracts of the ocean floor and bringing up the nodules. Technologists may be able to predict with fair accuracy the time and effort needed to bring this about despite the fact that they do not know how to do it today. The point is that all research is goal-oriented. Productivity increases occur not only through expenditure on research and development. It increases also through the process of resource-use, what one would call learning-by-doing, or learning-by-using. Getting children to learn their multiplication tables is proto-typical learning-by-doing: repeated use, in this case the arithmetic operation, teaches one to operate more efficiently. Past experience is often a reliable guide to the extent of future learning through doing, or through using.

We thus see that there is, to put it metaphorically, a constant tension between forces which *raise* extraction and refining costs – as high-grade deposits are depleted – and those which *reduce* such costs – as newer and newer technological processes are discovered. It is the balance between these two forces that determines the evolution of extraction and refining costs. Both are influenced by policies affecting extraction rates and R&D expenditure. As we will see shortly, unit costs of extraction and refinement of exhaustible resources, in the aggregate, fell steadily over the first 40 years of this century. But there are reasons for thinking that they have been rising slowly during the past few decades. We will study the implications of this.

Technological improvements and their influence on resource substitutability are a key to understanding the economics of exhaustible resources. One can therefore see why it is often suggested that future generations need increased knowledge and capital equipment to exploit currently *un*usable resources rather than just vast quantities of currently usable resources. But for technological discoveries to occur there needs to be a motive and thence an incentive. If serendipity is the drug which induces new discoveries, potential profits provide the far more reliable carrot and stick. These profits can be private or social, depending on the agency instituting the research – the lone inventor, the giant corporation, the private consortium, the government, or whatever. As we may infer from our seven-fold list of discoveries, these profits may be generated by the discovery of new deposits or from discovering ways of using existing deposits more efficiently. In short, there are profits to be had from expanding the resource base. As we are all painfully aware, research and development involves the expenditure of resources; not only by way of equipment but also in the form of salaries. In order to strike the appropriate balance in scientific and technological research the expected costs of research and development must be compared with the expected benefits (or profits) to be enjoyed from it. It is in the estimation of expected benefits that one needs to know the resource base and its economic characteristics.

It is essential to realise that it is the size of the existing usable stock and not the current resource price which influences the expected profitability of a new technological discovery. As we noted earlier, the four-fold price increase of Arab crude oil in late 1973 was not a signal for impending resource depletion. It was a message that a cartel was exercising its market power. Nevertheless, one might think that as a resource gets depleted its price rises, signalling growing scarcity. Thus, one might think that, ignoring such exceptions as resource cartels coalescing effectively, price increases are the signals the market provides for benefit-seeking inventors, firms and governments. It transpires that this is not quite precise enough. The matter is not unduly difficult to fathom, but it is a good deal deeper. So I now turn to the behaviour of market prices of exhaustible natural resources, to see *which* price we ought to be monitoring, and why.

Analysis of resource prices

A pool of oil, or a vein of iron or a deposit of coal in the ground is a capital asset to society – and to its owner if the deposit is owned by an individual agency. The deposit draws its value, say a market value, from the prospect of extraction and sale. In the meanwhile, as a distinguished economist aptly remarked, the resource owner is looking down at his mine and asking, 'What have you done for me lately?' There are two ways an asset can yield a return to its owner: by appreciating in value – capital gains – and by earning a dividend. Most assets offer a combination of both. (Of course, assets are all too often known to depreciate in value – capital losses – but this is merely the negative of appreciation and so falls in the same category.) The fundamental point about an exhaustible resource deposit lying untapped is that the only way it can earn a return for its owner is by appreciating in value. It cannot earn any dividend, by definition. We must now ask by how much the deposit should appreciate in value over time in order that investors are willing to hold a mixed portfolio consisting of shares in this deposit and shares of other market assets. This is easy to answer. In order for people to wish to hold a mixed portfolio of assets – inclusive of the resource deposit – the expected capital gains on the resource deposit must equal the expected rate of return on other capital assets in the same risk class. This is the fundamental principle of the economics of exhaustible resources and is today referred to by economists as the Hotelling Rule, in honour of the man who first deduced it in a rigorous manner while analysing the eighth mechanism in my earlier list.

Now, the capital gains I have been talking about are the rates of appreciation in the value of the resource deposit underground, the rate of capital gains on the value of the mine. They are thus the rate of appreciation of the *ground rent*, or *royalty*, on the mine. The Hotelling Rule

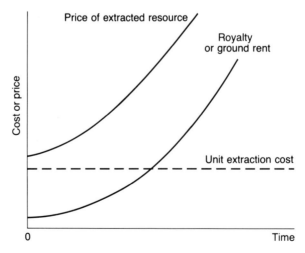

Fig. 6.1. Competitive price path of exhaustible resource: constant unit extraction cost.

states that the expected rate of capital gains on the mine will – under equilibrium conditions – grow in a compound manner at a rate r given by the expected rate of return on any other capital asset belonging to the same risk class. The royalty is *not* the final sales price of the extracted and refined ore. It is less, sometimes far less, because extraction and refinement, not to mention transportation, all involve costs. The final sales price of the extracted, refined and transported resource must be approximately the *sum* of its ground rent – or royalty – and its extraction, refinement and transportation cost.

The simplest general case to study is one where the cost of extraction of a unit of the ore is constant over time. The final sales price of the extracted ore is then a constant amount in excess of the royalty, which as we have seen, grows at a compound rate, r, equal to the rate of return on other assets. The final sales price of the refined ore in this case grows at a rate *less* than r. Figure. 6.1 displays the price trajectories of such a resource. Now, one would not expect extraction costs to remain constant, for reasons we have already discussed. If unit costs of extraction are expected to grow because society will be forced to mine more and more inferior grades, the final sales price will rise, for two reasons now, rather than the single reason prevailing if unit extraction costs remain constant. But if technological advances lower extraction costs at a sharp enough rate the final sales price would indeed fall. But it cannot fall forever, because extraction costs have a floor to them, namely, zero. Eventually the final sales price must rise, and rise at nearly a compound rate, for the ground rent term will eventually come to dominate (see Fig. 6.2). The key word here is 'eventually' and I shall come back to it.

If the existing, usable, deposit is large, the ground rent will initially be very low relative to extraction costs, but as the deposit eventually gets depleted the ground rent comes to dominate (Fig. 6.1).

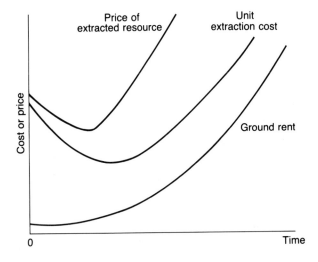

Fig. 6.2. Competitive price path of exhaustible resource: variable unit extraction cost.

This conclusion holds even if new deposits are continually being discovered through the process of search. It is an easy matter to check that the argument emanating from the fact that an untapped deposit cannot yield a dividend continues to hold even when new deposits are being discovered in a predictable way. The Hotelling Rule is thus robust against this form of generalisation.

Matters are different if there is uncertainty in the discovery process. In this case, prices that are expected are not necessarily realised. If discoveries over a period happen to be systematically larger than anticipated, the realised ground rent will decline over the period (see Dasgupta and Heal, 1975, 1979, and Arrow and Chang, 1980). Nevertheless, even in the face of uncertain discoveries, a modified version of Hotelling's Rule holds, the rule now pertaining to the ground rent which in some sense is *expected* to prevail, not to that which is ultimately realised.

Royalties as a measure of resource scarcity

Now, data on ground rents, or what one may call royalties, are hard to come by. However, there have been several analyses of time series of extraction costs and final sales prices of exhaustible resources. In their pioneering work titled *Scarcity and Growth*, Barnett and Morse (1963) argued that with the exception of timber, both extraction costs and final prices of natural resources displayed a secular decline during the first half of this century. If true, this must have been due to a combination of our first seven mechanisms. Barnett and Morse concluded from this that resources were not getting scarcer. On the contrary, they were getting less scarce.

Now this is an incorrect conclusion to draw. We have noted that if extraction costs fall fast enough the final sales price will fall for a while even if there were a single deposit and no discoveries in sight. We would scarcely say that the resource was in this case getting less scarce. So then which price signals growing scarcity? The ground rent of course: its 'expected' value grows at a compound rate, signalling the expected growing value of the resource. Thus ground rents, as a fraction of extraction costs, say, are a true index of resource scarcity. If the deposits are initially very large the ground rent is tiny and it is many years before the rent rises to appreciable figures. Even at 5% per annum – a generous figure for the real interest rate – a penny takes about 120 years to become a five-pound note. Compound interest is no doubt remorseless, but it can at the beginning be very slow if what is compounded is very small to start with.

In any case time series of resource prices are not unambiguous. Estimates of the behaviour of resource prices tend to vary from study to study. In a careful study Brown and Field (1979) assess different estimates and show that the prices of a selected group of key minerals did not display a secular decline during the first half of this century. Indeed, for one set of data (see Table 6.2) given for three selected years, prices display a U-shaped path.

Table 6.2. *Real price of selected minerals using the price of capital as a numeraire, selected years, 1920–50*

	1920	1940	1950
Coal	340	195	413
Copper	170	125	129
Iron	216	149	146
Phosphorus	—	141	170
Molybdenum	—	—	186
Lead	292	211	298
Zinc	301	281	335
Sulphur	—	—	281
Aluminium	647	297	217
Gold	—	615	337
Crude petroleum	547	205	278

Source: G. Brown and B. Field, 'The adequacy of measures for signalling the scarcity of natural resources. In V. Kerry Smith, *Scarcity and Growth Revisited*, Johns Hopkins University Press (Baltimore), 1979.

Transition from exhaustible to durable resource base

It should now be clear what I am leading up to. I began by suggesting that it is not specific resources we care for, it is certain characteristics we seek, certain services we need and desire. A wide variety of resources may offer the same set of services. This is precisely why *substitution* is the key concept in resource economics. Timber, coal, oil and natural gas are all sources of energy. So is uranium 238 a source of energy today, although it was not so 50 years ago. Moreover, if at some future date nuclear fusion comes to be controlled at an acceptable risk, or if solar energy comes to be tapped in large amounts at reasonable cost, we will have at our disposal vast new deposits of energy, so vast that to all intents and purposes we will have a renewable source, which is to say that the royalty on such sources, when they become technologically and economically viable for exploitation, will be tiny for a long while. But the economic availability of such renewable sources is at the moment conjectural. Provided resources are diverted towards research and development of these alternative sources we may expect to develop them at some time in the future, and the greater the R&D effort the earlier the expected date of success. Meanwhile we have to rely on finite resources. What then will be the price trajectory of the service provided by such resources? In Fig. 6.3, I present the competitive price path of a service, such as, say, energy, when it has to be obtained from a finite deposit initially and where it is expected that a renewable source will become available at a date, T, somewhat far away in the future. (For simplicity of exposition I assume away uncertainty once again, so that expectations are realised. See Dasgupta and Heal (1975) for an analysis of the case where T is uncertain.) Clearly, the expected date of completion of the R&D programme on the alternative energy source influences the current and future resource price. Clearly also, the size of the existing deposits of fossil

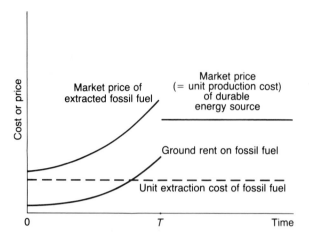

Fig. 6.3. T shows time of substitution of renewable energy source for non-renewable resources.

fuels will, and should, influence the rate of R&D effort and thus the expected date of completion. These mutual relationships can be analysed, and there is now a substantial body of literature which does precisely this, having taken into account such obvious extensions as uncertainty in research completion times and in the cost of energy obtainable from these future technologies (see Dasgupta, Heal and Majumdar, 1977, and Dasgupta, 1982).

Fig. 6.3, describing a possible time profile of energy price, is stylised to indicate the crucial features. The *transition* from the exhaustible, fossil fuel base to the more durable, renewable, nuclear or solar base will not be dramatic, unlike that shown in Fig. 6.3. If and when such renewable sources are put into operation alternate sources will co-exist, possibly for quite some time, as indeed they do now. Prices will not display the discontinuity suggested in Fig. 6.3 because the diagram has not allowed for the facts that new technologies inevitably suffer from teething trouble, that production costs go down as experience grows, and that plants take time to build. In short, transition from an exhaustible resource base to a durable, renewable, resource base can always be expected to be gradual.

The nagging question remains: isn't all this too fanciful? We began by discussing exhaustible resources and we are now making the mental transition to durable resource bases. Is this the right conceptual framework? The fact that economic theorists have analysed the implications of such possibilities provides no reason why this is the right way of going about thinking on the matter. In any case, I have been concentrating so far on energy resources. Why should we think this is the only biting constraint, the most important issue to discuss?

Prospects for transition to the age of substitutability

Such questions as these can only be broached by physical scientists, and as it happens two such people, H.E. Goeller and A.M. Weinberg of the Oak Ridge National Laboratory and the Institute for Energy Analysis at Oak Ridge, Tennessee, respectively, addressed precisely this class of questions in a bold and highly original article entitled 'The age of substitutability' (1976). To the best of my knowledge theirs is the first empirical attempt at this question. I will be surprised if their estimates, which I will present below, are not controversial. I will also argue that their estimates must not be taken literally, for they are cavalier in their treatment of some fundamental issues. Nevertheless, the spirit of their exercise is just right. For a first step this is what matters most.

Goeller and Weinberg went through the periodic table, examining all the elements and then some important compounds, such as hydrocarbons, noted their worldwide use in the year 1968 – the authors' base year – and estimated the global presence for each item, having leant on an extremely generous definition of potential resource deposits. The deposits included the atmosphere, the oceans and a mile-thick crust of the earth. Their attitude towards what constitutes a resource is light-years away from the attitude I have taken in this chapter: they make only cursory reference to costs of resource extraction and they concern themselves mainly with their physical *presence*, not with their economic *availability*. For example, a given quantity of a resource may be thinly distributed over the earth's crust, in which case extraction costs will be very high. Alternatively, the same amount may be present in highly concentrated forms, in which case extraction costs will be low. This, as we have noted, can make an enormous difference. (To illustrate the point I am driving at consider the fact that soil erosion – that is, loss of top soil – is today a major form of environmental degradation, a serious loss in world-wide agricultural output. The quantity of soil in the atmosphere, the seas and a mile-thick crust of the earth remains pretty much the same. But the soil has got 'relocated', it has got 'blown away', and this matters greatly.) Nevertheless, Goeller and Weinberg's treatment is just the kind of preliminary analysis one needs to complement the analysis that economists have provided, or if one feels more comfortable seeing it the other way, the economists' analysis is precisely the kind one needs to complement theirs.

I want first to give some idea of our current use of exhaustible materials. In Table 6.3 world demands in 1974 of the major non-renewable materials are presented. In terms of quantity, far and away the most important are sand, stone and related products (accounting for nearly 60%), and this is followed by fossil fuels (nearly 37%). These materials, plus iron, aluminium and magnesium, and eight of the other most widely used elements constitute more than 99.5% of all non-renewable materials currently used by society. If one includes prices and computes their relative importance in value terms, fossil fuels leap to first place, accounting for nearly 55% (right-hand column), building materials plummet to a rock-bottom place, accounting for under 5%,

Table 6.3. *World demand for non-renewable materials, 1974*

	World total demand	
Material	% of total quantity	% of total value
Fossil fuels (CH_x)		
Coal and peat	17.02	11.37
Petroleum	14.56	40.41
Natural gas	4.87	2.75
Subtotal	36.45	54.53
Sand and gravel (SiO_2)	32.28	1.98
Crushed stone ($CaCO_3$)	23.58	1.95
Clay, gypsum, pumice	2.94	0.75
Subtotal	58.80	4.68
Iron	2.63	24.44
Aluminium and magnesium	0.08	2.31
Subtotal	2.71	26.75
Other major		
Chlorine	0.14	0.66
Sodium	0.36	0.49
Nitrogen	0.37	1.91
Sulphur	0.26	0.27
Oxygen	0.26	0.16
Hydrogen	0.10	0.98
Potassium	0.10	0.23
(Phosphorus)	0.07	0.24
Subtotal	1.66	4.94
All other	0.38	9.10

Source: H.E. Goeller, 'The age of substitutability: a scientific appraisal of natural resource adequacy.' In *Scarcity and Growth Reconsidered*, V. Kerry Smith (ed.), 1979.

and iron, aluminium and magnesium jump to a second place at nearly 27%. In value terms these principal materials account for over 90% of all non-renewable materials currently used by society.

I now come to the Goeller–Weinberg estimates. In Table 6.4 data on the hydrocarbons and a few of the more notable elements are provided. The most revealing column is the last one. It presents, for each resource, the total stock in the atmosphere, the oceans and a mile-thick crust of the earth, divided by the 1968 world demand for it. In other words, it presents the number of years each resource can be expected to last at the 1968 rate of world use. (The underlying assumption is that recycling is not possible.) Now, demand must surely be expected to grow, though not necessarily at the same rate for all resources, as incomes rise and population grows, and I will come back to the adjustments that will be required to account for such changes. But note first that the only causes for worry are the phosphates – a mere 1300 years' supply – fossil fuels, labelled CH_x – some 2500 years – and manganese – about 13000 years. The rest are available for more than a million years, which is pretty much like being inexhaustible.

At a more specific level Goeller and Weinberg argue that on the basis of geological and technological data, the essential raw materials are effectively in infinite supply, with the exception of phosphorus, fossil fuels and some trace elements needed for agriculture, notably zinc and copper. They argue that as the world exhausts its non-renewable raw materials it can move to lower-grade inexhaustible resources. The increase in the energy required to extract and process these low-grade materials is not huge. Goeller and Weinberg present crude estimates of the ratio of the energy required to extract a ton of some of these low-grade abundant metal ores to that required to extract a ton from high-grade ores. In none of these cases does the ratio exceed two. (See Table 6.5. One should note though that these *are* crude estimates and, as I mentioned earlier, not much attention has been given to the concentration in which the raw materials are found in the earth's crust.)

Table 6.4. *Present or future nearly inexhaustible resources for the most extensively used elements. The last column gives R/D, the resource-to-demand ratio in 1968.*

Element	Resource	World resource (tons)	R/D (years)
CH_x (extractable)	Coal, oil, gas	1×10^{13}	2500
C (oxidised)	Limestone	2×10^{15}	4×10^6
Si	Sand, sandstone	1.2×10^{16}	5×10^6
Ca	Limestone	5×10^{15}	4×10^6
H	Water	1.7×10^{17}	$\sim 10^{10}$
Fe	Basalt, laterite	1.8×10^{15}	4.5×10^6
N	Air	4.5×10^{15}	1×10^8
Na	Rock salt, seawater	1.6×10^{16}	3×10^8
O	Air	1.1×10^{15}	3.5×10^7
S	Gypsum, seawater	1.1×10^{15}	3×10^7
Cl	Rock salt, seawater	2.9×10^{16}	4×10^8
P	Phosphate rock	1.6×10^{10}	1300
K	Sylvite, seawater	5.7×10^{14}	4×10^7
Al	Clay (kaolin)	1.7×10^{15}	2×10^8
Mg	Seawater	2×10^{15}	4×10^8
Mn	Sea floor nodules	1×10^{11}	13000
Ar	Air	5×10^{13}	2×10^8
Br	Seawater	1×10^{14}	6×10^8
Ni	Peridotite	6×10^{11}	1.4×10^6

Source: H.E. Goeller and A.M. Weinberg, 'The age of substitutability', *Science*, 20 February 1976.

Table 6.5. *Energy requirements for the production of abundant metals and copper. Gross energy is estimated at 40% thermal efficiency for generation of electricity. The last column gives the ratio of the energy required to extract a ton of metal for low-grade (E_L) compared to high-grade (E_H) ores*

Metal	Source	Gross energy (kilowatt-hours per ton of metal)	E_L/E_H
Magnesium ingot	Seawater	100 000	1
Aluminium ingot	Bauxite	56 000	1
	Clay	72 600	1.28
Raw steel	Magnetic taconites	10 100	1
	Iron laterites	11 900	1.17 (with carbon)
			~2 (with electrolytic hydrogen)
Titanium ingot	Rutile	138 900	1
	Ilmenite	164 700	1.18
	Titanium-rich soils	227 000	1.63
Refined copper	Porphyry ore, 1% Cu	14 000	1
	Porphyry ore, 0.3% Cu	27 300	1.95

Source: E.H. Goeller and A.M. Weinberg, 'The age of substitutability', *Science*, 20 February 1976.

Thus the problematic resources are phosphorus, some trace elements needed for agriculture, and fossil fuels. The solution for phosphorus and the trace elements must then lie in recycling. Indeed, H.G. Wells, in his book *The World Set Free*, had already suggested that phosphorus is in limited supply and that we may need to recycle it as fertiliser by crushing bones! The supply of hydrocarbons, adjusting for population growth, which Goeller and Weinberg take as 10 billion in the long run, will last only a few hundred years, a good deal less than the 2500 years estimated on the basis of 1968 demand. This then is the fly in the ointment, the bottleneck, the binding resource: limited hydrocarbons, currently the main source of energy. In a sense we are back, full circle, to where we began in this chapter, the exhaustibility of energy resources. Here Goeller and Weinberg have an interesting suggestion. They observe that both carbon and hydrogen are abundant in oxidised form, specifically, in the form of carbon dioxide and water respectively. They also note that there are a number of known technical processes which allow one to reduce carbon dioxide and water to hydrocarbons needed for transport and energy. The only trouble is, these processes in turn require a good deal of energy! So there we are, back to the 'ultimate' resource: cheap and non-risky sources of usable energy. In an interesting work *The Next Million Years*, C.G. Darwin, grandson of the man round whose works this Lecture Series revolves, saw only one possible source of unlimited energy: nuclear fusion.

Today we conceive of other inexhaustible sources, solar, geothermal and clean nuclear breeder reactors.

Goeller and Weinberg see the pattern of future resource-use as consisting of three stages. The first, lasting for another 30–40 years at least, would be a continuation of present patterns of use of exhaustible resources. The second, lasting several hundred years, would be one where society depends mainly on coal – there would be little oil and natural gas left – and less reliance would be placed on non-ferrous metals, more on alloy steels, aluminium, magnesium and titanium. Beyond that is the third stage, the Age of Substitutability, by which time fossil fuels will have been pretty much exhausted. In this final age economic activities will be based almost exclusively on materials that are virtually inexhaustible, with relatively little loss in living standards. Goeller and Weinberg speculate that the Age of Substitutability will be based largely on glass, plastics, wood, cement, iron, aluminium and magnesium.

Such a situation may be difficult to imagine perhaps, but then, eighteenth-century society would have found it difficult to imagine us today, certainly they would be astonished at the resource base upon which modern economic activity rests. But for me the importance of the Goeller–Weinberg study lies in that they asked the right questions and they attempted to chart the right avenue for locating answers to these questions. As a start, one must try and account for what resources are around us and what technological processes are currently known or are in the offing. It is only against such a background that one can begin to discuss policy options, such as, for example, the relative merits of different research and development programmes. The details of the Goeller-Weinberg speculations are not of great importance, nor would they, I imagine, wish to emphasise them. The merit of their work, in contrast to the far more well-known and well-packaged World Models of Jay Forrester, Dennis Meadows and their collaborators, is that Goeller and Weinberg have attempted only to get things approximately right, not accurately wrong.

And yet, talk of controlled nuclear fusion and clean breeder reactors smacks of futurology, wishful thinking. If such technological possibilities appear to be remote now, and this remoteness a cause for anxiety about the future, we should remember that even in the second half of the *nineteenth* century thoughtful people worried about declining energy sources and saw no way out. In a famous chapter on possible substitutes for coal, the economist W.S. Jevons, in his book *The Coal Question*, published in 1865, expressed deep concern about the future of British industry. And he wrote:

> I draw the concluson that I think anyone would draw, that we cannot long maintain our present rate of increase of consumption; that we can never advance to the higher amounts of consumption supposed . . . the check to our progress must become perceptible considerably within a century from the present time; that the cost of fuel must rise, perhaps within a lifetime, to a rate threatening our commercial and manufacturing supremacy; and the conclusion is inevitable; that our happy progressive condition is a thing of limited duration.

That Britain's commercial and manufacturing supremacy was under threat cannot be doubted, since the supremacy did not last. But the price of fuel had nothing to do with it.

Well, I can almost *hear* the reader thinking, trust an economist to goof so badly. But before economists are judged to be systematically worse at this sort of thing than others I will quote a passage from an editorial comment in the September 1876 issue of the *Scientific American*:

> A recent lecture was given by Professor Carey Foster, F.R.S., at South Kensington on 'Electricity as a Mode of Power'. Dr Siemens, F.R.S., took the chair. Professor Foster first showed with a number of experiments that by very simple arrangements light bodies can be moved by electricity . . . Although we cannot say what remains to be invented, we can say that there seems no reason to believe that electricity will be used as a practical mode of power. There is always power lost by the inverse current. Work of some kind must be done to produce electricity, and this can more economically be done by employing that power directly.

The moral is banal. We are all given to myopia.

Further reading

There is a large historical literature on resource substitution. The sample below is as interesting as any I know:

Rosenberg, N. (1976). *Perspectives on Technology*, Cambridge University Press, Chapters 13 and 14.

Rosenberg, N. (1980). Technology, natural resources and economic growth. In Bliss, C.J. & Boserup, M. (eds.). *Economic Growth and Resources*, Vol. 3: *Natural Resources*. London: Macmillan.

Scott, A. (1962). The development of extractive industries, *Canadian Journal of Economics*, **28**, 70–87.

Thomas, B. (1980). Towards an Energy Interpretation of the Industrial Revolution, *Atlantic Economic Review*, **8**, 1–15.

The decade of the 1970s saw vigorous activity in the economics of exhaustible resources, both analytical and empirical. The earliest mathematical treatment of the subject is:

Hotelling, H. (1931). The Economics of Exhaustible Resources, *Journal of Political Economy*, **39**, 137–75.

A characteristically lucid treatment of Hotelling's analysis and its extensions completed by 1973 can be found in:

Solow, R.M. (1974). The economics of resources, or the resources of economics, *American Economic Review*, **64** (Papers and Proceedings), 1–21.

The development of a unified framework for analysing various resource substitution mechanisms was undertaken in a series of articles by a number of economists:

Dasgupta, P. & Heal, G. (1975). The optimal depletion of exhaustible resources. *Review of Economic Studies,* Symposium Issue, pp. 1–23.

Dasgupta, P., Heal, G. & Majumdar, M. (1977). Resource depletion and research and development. In M. Intriligator (ed.), *Frontiers of Quantitative Economics,* Vol. III. Amsterdam: North Holland.

Dasgupta, P., Gilbert, R., & Stiglitz, J. (1982). Invention and innovation under alternative market structures: the case of natural resources, *Review of Economic Studies,* **49,** 567–82.

Dasgupta, P., Gilbert, R., & Stiglitz, J. (1983). Strategic considerations in invention and innovation: the case of natural resources, *Econometrica,* **51,** 1439–48.

Dasgupta, P. & Stiglitz, J. (1981). Resource depletion under technological uncertainty, *Econometrica,* **49,** 85–104.

Gallini, N., Lewis, T. & Ware, R. (1983). Strategic timing and pricing of a substitute in a cartelized resource market, *Canadian Journal of Economics,* **16,** 429–46.

Arrow, K.J. & Chang, S. (1980). Optimal pricing, use, and exploration of uncertain natural resource stocks. In Liu, P.T. (ed.), *Dynamic Optimization and Mathematical Economics.* New York: Plenum Press.

Hoel, M. (1978). Resource extraction, substitute production and monopoly, *Journal of Economic Theory,* **19,** 28–37.

Semi-technical presentations of this literature can be found in:

Dasgupta, P. (1981). Resource pricing and technological innovations under oligopoly: a theoretical exploration, *Scandinavian Journal of Economics,* **83,** 289–318.

Dasgupta, P. (1982). Resource depletion, research and development and the social rate of return. In R. Lind (ed.), *Discounting for Time and Risk in Energy Policy.* Baltimore: Johns Hopkins University Press.

An outstanding introductory text on the economics of natural resources is:

Hartwick, J.M. & Oleweiler, N.D. (1986). *The Economics of Natural Resource Use.* New York: Harper and Row.

For an advanced treatise on the subject, see:

Dasgupta, P. & Heal, G. (1979). *Economic Theory and Exhaustible Resources.* Cambridge University Press.

Estimates of the world's resource deposits, the behaviour of resource prices and discussions of measures of resource scarcity can be found in:

Barnett, H.J. & Morse, C. (1963). *Scarcity and Growth: The Economics of Natural Resource Scarcity.* Baltimore: Johns Hopkins University Press.

Goeller, H.E. (1979). The age of substitutability: a scientific appraisal of natural resource scarcity. In Kerry Smith, V. (ed.), *Scarcity and Growth Reconsidered*. Baltimore: Johns Hopkins University Press.

Goeller, H.E. & Weinberg, A.M. (1976). The age of substitutability, *Science*, **191** (February 20), 683–9.

Harris, D.P. & Skinner, B.J. (1982). The assessment of long-term supplies of minerals. In Kerry Smith, V. & Krutilla, J.V. (eds.), *Explorations in Natural Resource Economics*. Baltimore: Johns Hopkins University Press.

Zimmerman, E.W. (1964). *Introduction to World Resources*. New York: Harper and Row

Brown, G. & Field, B. (1979). The adequacy of measures for signalling the scarcity of natural resources. In Kerry Smith, V. (ed.), *Scarcity and Growth Revisited*. Baltimore: Johns Hopkins University Press.

Slade, M.E. (1982). Trends in natural-resource commodity prices: an analysis of the time domain, *Journal of Environmental Economics and Management*, **9**, 122–37.

Debrajan, S. & Fisher, A.C. (1982). Measures of natural resource scarcity under uncertainty. In Kerry Smith, V. & Krutilla, J.V. (eds.), *Explorations in Natural Resource Economics*. Baltimore: Johns Hopkins University Press.

For a study of, among other things, the effect of recent increases in the price of oil on the world's economy see:

Bruno, M. & Sachs, J. (1985). *The Economics of Worldwide Stagflation*. Oxford: Basil Blackwell.

An outstanding treatment of the transition problem, that is, the transition from exhaustible to inexhaustible resource bases, in which both analytic and numerical issues are highlighted, can be found in:

Koopmans, T.C. (1980). The transition from exhaustible to renewable or inexhaustible resources. In Bliss C. & Boserup, M. (eds.), *Economic Growth and Resources, Vol 3: Natural Resources*. London: Macmillan.

For a treatment which highlights the substitution of fixed capital for vanishing resource-use in production, see:

Stiglitz, J. (1979). A neoclassical analysis of the economics of natural resources. In Kerry Smith, V. (ed.), *Scarcity and Growth Reconsidered*. Baltimore: Johns Hopkins University Press.

[7]

Changing climates

Bert Bolin

Climatic changes in the past

The climate on earth has changed throughout the evolution of our planetary system over about 4500 million years. Sometimes these changes have been rapid. During other periods they have been slow, hardly noticeable over millions of years, as far as we can judge from the evidence that is found to-day in rocks and sediments. To be able to discuss possible future climates we must understand the mechanisms behind past changes. The climatic system is too complex for us to rely merely on simple extrapolation of observed trends and variations or theoretical models that have not been tested against past data.

We know the gross features of climatic change on earth during the last major geological epochs, i.e. during the last few hundred million years. These have been determined by analyses of the abundance of the oxygen-18 isotope in remains of planktonic micro-organisms (foraminifera). The ratio of oxygen-18 to the normal isotope, oxygen-16, reflects water temperature and global ice volume. High proportions of oxygen-18 correspond to glacial periods when more of the lighter isotope oxygen-16 was locked up in ice. These analyses reveal that water temperatures in the deep layers of the oceans have been close to zero degrees only during the last few million years, i.e. the Quaternary period, and also that ocean surface temperatures were probably higher during these earlier periods than they are today (Fig. 7.1). Presumably the climate on land was also significantly different from to-day.

A much more detailed record of the proportion of oxygen present as oxygen-18 for the Quaternary period shows marked variations on much shorter timescales. During the last 350 000 years (Fig. 7.2), which covers just a few thousandths of the time period depicted in Fig. 7.1, we find quasiperiodic variations on a timescale of about 100 000 years. We notice particularly the change from a comparatively small oxygen-18 value about 120 000 years ago through significantly larger values during most of the last 100 000 years, back again to smaller values during the last 12 000 years. The low values are generally interpreted as corresponding to periods of minimum glaciation, 120 000 years ago as well as at present, and the high values as corresponding to markedly more ice locked into continental ice sheets during the inter-mediate period. Thus, the last Ice Age, with glaciation of regions that to-day

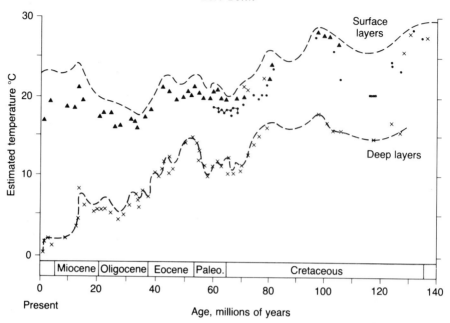

Fig. 7.1. Temperature estimates based on oxygen – 18 measurements
in foraminifera from deep ocean layers and surface layers of the ocean
water at low latitudes during the last 140 million years. (From Douglas
& Woodruff, 1981)

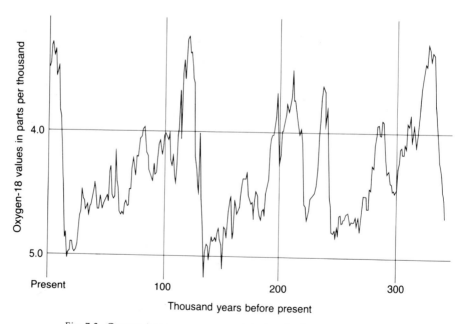

Fig. 7.2. Oxygen isotope measurement during the last 350 000 years.
Note that the top of the graph represents higher temperatures. (From
Shackleton & Pisias, 1985)

experience a rather mild climate, seems to have lasted for about 100 000 years, although it was not necessarily uninterrupted during this whole period.

Even more detailed series of observations are available for the last 15 000 years, i.e. the deglaciation phase and the interglacial that followed and that we still are enjoying. We find during this period climatic variations on even shorter timescales, from a few hundred years to a few thousand years. For example, the oscillation during the period 11 000 – 10 000 years ago represents a temporary halt in the retreat of the glaciers that seems to have lasted for about 500 years. The warm climatic conditions then returned, which resulted in rapid decrease of both the ice-sheet over Scandinavia and the Laurentide ice-sheet over eastern Canada. A comparatively warm period, probably somewhat warmer than today, at least in north-western Europe, prevailed about 6000 years ago, which, however, slowly gave away to a somewhat cooler climate. There were comparatively mild conditions about 1000 years ago, when Iceland was colonised by the Norwegians, some settlements were established in southern Greenland and even the north-eastern most part of America was discovered by the 'North men'; but with this exception the slow deterioration of climate at middle and high latitudes continued. We know that cultivation of wine had to be abandoned in England in the fourteenth century and that winters were cold in Europe during the sixteenth and seventeenth centuries as compared with today, with the advance of glaciers in Scandinavia and the Alps. Seldom do we find winter conditions to-day in Holland comparable with those pictured by Flemish painters during this time. During the last 100–200 years the climate on earth has again become warmer.

Regular temperature measurements have been made at a few stations for little more than 200 years, but enough data to establish changes of the global mean temperature more accurately are available for only about 120 years. The most careful analysis of these data has been carried out at the Climate Research Unit of the University of East Anglia. Fig. 7.3 shows their most recent results. An upward trend is clearly visible and the rate has been + 0.5± 0.2°C during the last century. The uncertainty stems from three sources: the uncertainty of data early in this period, the fact that few data are available in oceanic areas, particularly in the southern hemisphere, and the variations of the annual mean value on a timescale of decades which can be seen in the record. These circumstances preclude a more accurate determination of the long-term trend so far.

There have also been marked regional variations of climate during this period of time. It became much warmer in the Spitsbergen region during the first half of this century and of course catastophic droughts have recently hit parts of the African continent, particularly the Sahel zone and Ethiopia.

Some causes for climatic change

The climate variations described above show that changes have occurred on many timescales and that presumably they were determined by different

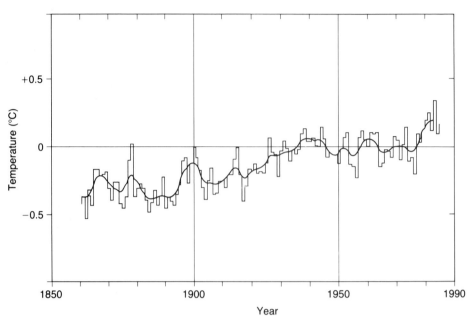

Fig. 7.3. Global mean surface temperature during the last 120 years.
(From Wigley et al., 1986)

causal factors. The changes during periods of hundreds of millions of years probably depend primarily on the drift of the continents, i.e. changes of the distribution of land and sea and of the topographic features of the earth's surface. On the other hand, it is remarkable, as Lovelock (1979) has pointed out, that there does not seem to have been any long-term trend towards general warming of the earth in spite of the increase of the intensity of the solar radiation that has probably occurred throughout the history of the earth.

The variations on the timescales of changes from a few hundred thousand to ten thousand years (Fig. 7.2) which seem to have persisted during the last 2.5 million years have most likely been caused by quasi-periodic changes of the earth's orbit around the sun. This would result in periodic anomalies in the distribution of solar radiation. For example, about 10 000 years ago the northern hemisphere experienced markedly warmer summers and colder winters than at present.

Computations using climate models confirm the role of these variations of the earth's orbit around the sun and associated variations of solar radiation in starting the retreat of the major ice sheets about 12 000 years ago. It should be emphasised, however, that another important factor probably has been the positive feedback due to the associated variations of the area covered by ice resulting in a change of the earth's albedo (reflectivity). During the 12 000 years that have elapsed since deglaciation began, the seasonal and latitudinal distribution of solar radiation has again changed back to glacial conditions. Expanding areas of snow and ice during winter and spring might trigger another ice age in the not too distant future. We do not know, however, when this might occur. The uncertainty of a projection on the basis of past changes is at least of the order of several thousand years, or about 100 generations of the human population. It is hardly meaningful to consider the possibility of another glaciation in planning for the future.

The reasons for changes of climate on timescales of hundreds to a few thousands of years, on the other hand, are not well understood. Satellite observations of the total flux of solar energy during the last decade indicate a decrease by as much as 0.1% in a decade. It is not known whether this is a short-term variation or a more long-term change which might prevail for a good part of a century. In the latter case such variations might well be reflected in the temperature variations that have occurred during the time period covered by reasonably good global observations (Fig. 7.3). As yet we have no basis for prediction of such climatic changes in the future. Still, they will presumably occur as they have in the past. The rise of temperature during this last century might in part be a natural variation of the earth's climate.

Today, however, we are confronted with another question. Are the ever-increasing human activities on earth changing the climate, and how rapidly might such a change take place? As we shall see, we are confronted with processes on timescales from decades to a few centuries, i.e. processes that are ten to a hundred times more rapid than the possible slow drift of the climate towards another ice age. It is most important to keep in mind this difference in the characteristic timescales of the processes on earth, which are important in determining the climate and its change.

The composition of the atmosphere and its role in the balance of radiation

The temperature at the earth's surface represents a balance between the incoming solar radiation on one hand and the loss of energy to space due to infra-red (IR) radiation on the other. We can compute the expected mean temperature at the earth's surface with the present intensity of solar radiation and albedo of the earth's surface but without an atmosphere. The result is about 254 K (i.e. $-19°C$). This is 35°C below the current condition. The much more favourable climate that we enjoy is due to the presence of water vapour, carbon dioxide, ozone and some minor constituents in the earth's atmosphere. Because they absorb infra-red, these gases prevent more than half of the radiation that leaves the surface of the earth from escaping to space directly. Instead, some of this absorbed energy is returned back to earth. This so-called 'greenhouse' effect is primarily due to water vapour and to a lesser extent to carbon dioxide and other minor consitituents. It is obviously of fundamental importance for the earth to be habitable. Today Man is changing these radiative characteristics of the atmosphere. We are increasing the concentration of carbon dioxide, next to water vapour the most important 'greenhouse' gas in the atmosphere, and we are also adding other gases to the atmosphere that affect the radiation balance of the earth's surface in a similar way.

Changing concentrations of atmospheric carbon dioxide

The Swedish chemist Svante Arrhenius (1896) had already pointed out before the end of last century that an increasing amount of carbon dioxide in the atmosphere might cause a warmer climate. This idea has remained alive during the twentieth century. Callender (1938) showed rather convincingly in the 1930s that the atmospheric CO_2 concentration was increasing. We now have 30 years of continuous measurements, principally due to C.D. Keeling. The observations from Mauna Loa on Hawaii shown in Fig. 7.4 have become classical evidence of man's global influence on the composition of the earth's atmosphere. The average annual *increase* has accelerated from about 0.7 ppmv (parts per million of volume) per year in 1957 to 1.3 ppmv per year at present. Superimposed on this general trend we see marked seasonal oscillations, which are more pronounced in North Polar regions, but hardly noticeable in Antarctica. The maximum concentration (in the northern hemisphere) is reached in April/May, the minimum in September. This oscillation is due to the seasonal variation in photosynthesis and in organic matter decay in the soils of the terrestrial ecosystems in the northern hemisphere.

Due to ingenious work by Swiss and French scientists, we now also know the variations of atmospheric carbon dioxide during the last 160 000 years. Glacier ice is formed from snow, which gradually becomes packed by subsequent layers of snow. During this process, small bubbles of air are enclosed in the ice and these represent an invaluable archive of air samples, which can be dated quite accurately. Fig. 7.5 shows analyses for the last few

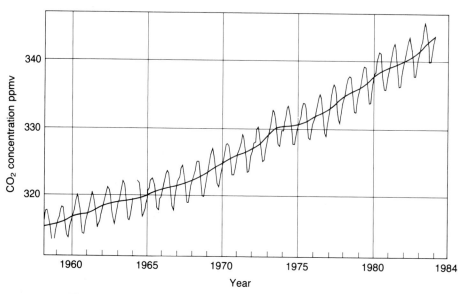

Fig. 7.4. Variations of atmospheric CO_2 at Mauna Loa Observatory, Hawaii. (Data reported by Bacastow & Keeling, 1981, supplemented by data from recent years supplied by personal communication.)

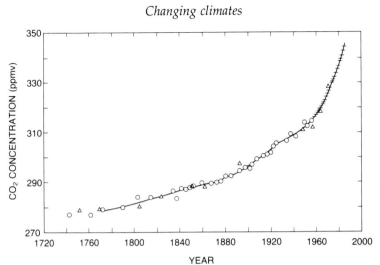

Fig. 7.5. Atmospheric CO_2 increase during the last 200 years as indicated by measurements in old ice from Siple station, Antarctica (triangles), and by gas chromatography (circles) and by annual mean values from Mauna Loa Observatory (Siegenthaler and Oeschger, 1987).

hundred years and yields a most important piece of information, namely that the pre-industrial CO_2 concentration in the atmosphere was 280 ± 5 ppm (Neftel *et al*, 1985). The increase during the industrial era since, say, the beginning of the nineteenth century, has been 70 ± 5 ppm, which is equivalent to $(148 \pm 11) \times 10^9$ t of carbon. The analysis of deeper strata in the Greenland and Antarctic ice sheets shows that the atmospheric CO_2 concentration may have been as low as about 200 ppmv towards the end of the last glaciation, i.e. about 15 000 years ago.

The change in recent years is undoubtedly due to human activities. Man is burning oil, gas and coal (Fig. 7.6). Forests are reduced and, although parts of the timber go into long-lasting structures, most of the wood is burned within a few years. Furthermore, the cultivation of soils unavoidably causes a decrease of soil organic matter due to an enhanced rate of decomposition. Estimates yield a total emission of carbon in the form of CO_2 by burning fossil fuels since 1860 at $(190 \pm 12) \times 10^9$ t of carbon (Rotty and Marland, 1986) and a decrease of carbon stored in living and dead organic matter on land over the same period of $(150 \pm 50) \times 10^9$ t of carbon (Bolin, 1986). Thus, an estimated $(340 \pm 62) \times 10^9$ t of carbon have been emitted as CO_2 over the industrial era, but we observe a rise of CO_2 in the atmosphere equivalent to only $(148 \pm 11) \times 10^9$ t of carbon. Less than half $(45 \pm 10\%)$ of the CO_2 emitted into the atmosphere has remained there. We naturally ask the question: Where has the rest gone?

To answer this question we need to analyse the pathways of carbon in nature, i.e. the carbon cycle. Since carbon is the fundamental element of life, the global carbon cycle also describes the basic global characteristics of both the terrestrial and marine ecosystems. Although a detailed study of these very complex systems is necessary before we truly understand their

[133]

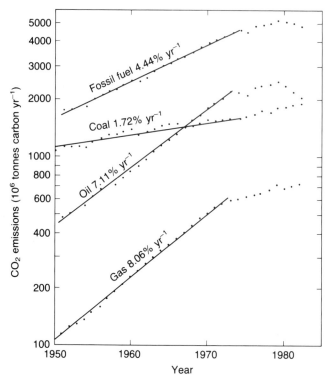

Fig. 7.6. Annual CO_2 emissions resulting from fossil fuel combustion 1950–82. (From Rotty & Masters, 1984)

behaviour, an overall picture of the major carbon reservoirs in nature and their large-scale exchange of carbon is helpful in the present context. Fig. 7.7 shows the gross features of interest. We note that the amount of carbon in living matter on land, 560×10^9 t, is merely about 75% of the amount presently found in the atmosphere, 735×10^9 t, while on the other hand the amount of carbon in the form of coal, oil and gas within possible reach, although not commercially exploitable at present (see Dasgupta, this volume) is 10–20 times larger than the amount temporarily stored in terrestrial biota. We conclude that continued exploitation of the world forests would contribute to a modest increase of atmospheric CO_2, while the use of future fossil fuels might cause a major change.

The role of the oceans in the global carbon cycle as the major sink for the emissions of CO_2 by human activities is crucial in the present context. Fig. 7.7 shows that there is about 50 times more carbon dissolved in the oceans than is present in the atmosphere to-day. In the surface layers of the oceans, where light is available, carbon is used by phytoplankton in the process of photosynthesis. These in turn serve as food for zooplankton and thereby constitute the base of the marine foodweb. There is a constant flux of dead organic matter in particulate form from the surface layers of the oceans into the deep sea. During this transfer the organic compounds are decomposed by

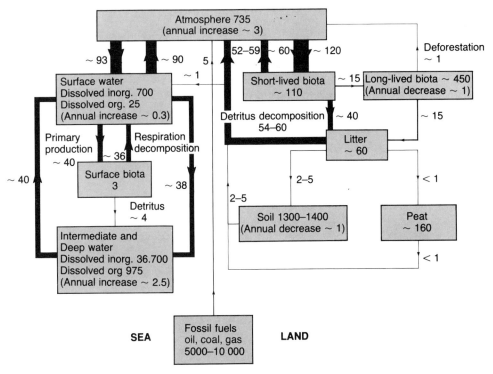

Fig. 7.7. The carbon cycle describing the major carbon reservoirs in
nature and their interchange of carbon. Reservoir sizes in units of $10^9 t$
of carbon and fluxes between reservoirs in units of $10^9 t$ of carbon per
year. (From Bolin, 1986)

bacteria and gradually the carbonates in the form of shells and other
structures are dissolved due to increasing pressure at increasing depth.
Accordingly there is a constant flux downwards of carbon and we find that
concentrations of dissolved inorganic carbon in the deep sea are about 15%
higher than in the surface layers. During pre-industrial conditions an
approximate balance prevailed between this downward flux of carbon by
settling particles on one hand and the upward transfer due to water exchange
between the carbon-rich deep sea and the surface layers on the other.
Photosynthesis in the sunlit surface waters is a major sink for CO_2. It accounts
for a reduction of partial pressure of CO_2 of more than 50% compared with
that at greater depths (350 ppmv compared with 800 ppmv). If photosynthesis
in the sea were to cease, the concentration of CO_2 in the atmosphere would
double.

We can estimate the rates of vertical water exchange in the sea by
measuring the amount of radioactive carbon, ^{14}C, present in different parts of
the oceans, in a similar way to the dating of archaeological objects that contain
carbon. The Atlantic Ocean seems to be turning over in about 200 years, while
the vertical water exchange in the Pacific Ocean may take about 1000 years.
These rates are slow compared to the rapid increase of CO_2 in the atmosphere

during the last half century. Although the oceans constitute a large carbon reservoir, and could accommodate a major part of the emissions caused by man, the deep waters are not rapidly accessible as a sink. Theoretical estimates indicate that probably less than half of the emissions so far have been transferred into the oceans. So we cannot yet account for the missing emitted CO_2 by merely considering the uptake by the oceans.

Several other carbon reservoirs may play a role in serving as sinks for man's emissions of CO_2 into the atmosphere. The eutrophication of lakes and coastal waters often also involves the incorporation of organic carbon into sediments. The enhanced amount of CO_2 in the atmosphere may have increased the rate of photosynthesis, and thus fixation of carbon in plant material, in those parts of the terrestrial ecosystem that man has not otherwise disturbed so far. In the deforestation process some of the organic matter may be left in the soils as elemental carbon (soot and charcoal) rather than being released into the atmosphere in the form of CO_2. The exchange of carbon with carbonate sediments may also possibly play a role in the global carbon cycle in a way that has not been recognised adequately so far.

How will the atmospheric carbon dioxide increase in the future?

We certainly know little of how the CO_2 released to the atmosphere may be partitioned between the different carbon reservoirs in nature, but most of the uncertainty in attempting to forecast likely future increases of atmospheric CO_2 is due to the fact that we cannot forecast future emission. The United Nations Environmental Program, the World Meteorological Organization and the International Council of Scientific Unions have made a joint effort to assess the likely future development. Upper and lower bounds of future emissions due to fossil fuel combustion have been estimated and the associated increase of atmospheric carbon dioxide concentrations derived (Fig 7.8). The upper bound is characterised by a four-fold increase of present emissions (which corresponds to an annual increase of emissions by about 2–2.5%) to about 20×10^9 t C year^{-1} in 2050. Higher values seem unlikely in view of environmental, social and logistic constraints. In this case a doubling of the preindustrial atmospheric CO_2 concentrations (560 ppmv) would be reached by the middle of next century. The lower bound is placed at 2×10^9 t C year^{-1} in 2050 and could possibly be achieved by sustained global efforts to limit the future use of fossil fuel energy. In this case, it seems likely that the airborne fraction of the emissions would decrease.

During the post-war period, from 1945 to 1973, the fossil fuel emissions of CO_2 into the atmosphere increased by 4.6% per year (see Fig. 7.6). The two oil crises in 1973 and 1979 have markedly changed this picture. During the first four years of the 1980s the emissions even decreased. Because of the slow response characteristics of the oceans we have not yet seen much of a reduction in the rate of increase of atmospheric concentrations, but this will probably become noticeable during the next decade if the earlier rapid increase of fossil fuel use is not resumed. In the case of a persistent reduction of the emissions to the lower-bound estimate, the increase of atmospheric CO_2 might almost come to a halt and not exceed 400 ppmv; that is, the natural sinks for CO_2 would account for 100% of emissions, rather than 55% as is the present case.

Changing concentrations of other greenhouse gases in the atmosphere

Not until the latter part of the 1970s and during the 1980s has it been fully recognised that man is changing the concentrations of a number of other radiatively active (greenhouse) gases in the atmosphere. We shall restrict ourselves here to considering a few of the most important ones, i.e. methane (CH_4), nitrous oxide (N_2O), ozone (O_3) and the chlorofluorocarbons (CFCs): $CFCl_3$ (freon 11) and CF_2Cl_2 (freon 12).

Adel (1939) observed strong atmospheric absorption in the infra-red region about half a century ago, which he attributed to the presence of *methane* in the atmosphere. In the late 1960s *in situ* measurements showed a methane concentration of 1.4 ppmv. During the last 15 years, a large number of

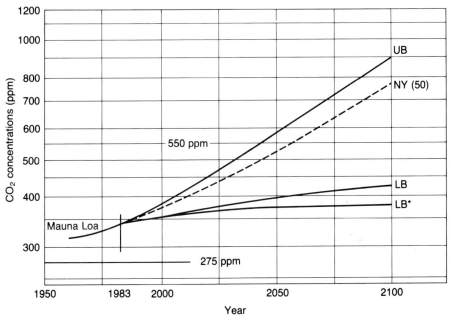

Fig. 7.8. Scenarios for future atmospheric CO_2 concentrations. UB (upper bound) corresponds to emissions from fossil fuel combustion of $20 \times 10^9 t$ and from terrestrial ecosystems of $2 \times 10^9 t$ of carbon per year, of which 50% remains in the atmosphere. LB (lower bound) corresponds to emissions from fossil fuels of $2 \times 10^9 t$ and from terrestrial sources of $10^9 t$ of carbon per year of which 40% remains in the atmosphere whereas LB* assumes enhanced uptake by ocean waters. Pre-industrial concentration (275 ppmv) and twice that value (550 ppmv) are also shown.

[137]

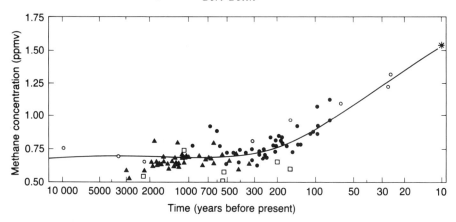

Fig. 7.9. Methane concentration measured in air trapped in ice cores as a function of time. Dots and filled triangles are data taken from Rasmussen & Khalil (1984) and represent values obtained from ice cores in Greenland and Antarctica respectively. Open circles are data published by Craig & Chou (1982), and squares are data published by Robbins *et al.* (1973).

measurements have been made which reveal a steady increase of the methane concentrations in the atmosphere by 1.1–1.3% per year to a global average value in 1985 of about 1.65 ppmv (Fig. 7.9). By analysis and dating of gas bubbles trapped in ice from the Greenland ice sheet, it has been possible to determine the methane concentration during preindustrial time to about 0.7 ppmv. Thus the amount of methane in the atmosphere has more than doubled during the last 200 years.

The natural amount of methane in the atmosphere represents a balance between two factors. On one hand, methane is produced by decomposition of organic matter under anaerobic conditions on land, in lakes and in the coastal zone of the oceans and, on the other hand, it is removed by photochemical oxidation caused by interaction with the radical OH in the atmosphere which, in turn, is produced by UV (ultra violet) radiation. The average residence time of a methane molecule in the atmosphere is about 10 years. The observed increase is possibly partly due to less OH radicals being available because of increasing air pollution in general, but it primarily depends on increasing emissions. These in turn are caused by increasing numbers of ruminants, whose gut micro-organisms produce methane, by expanding areas used for rice cultivation, burning of vegetation, and by leakage in the exploitation of natural gas and coal mining. We are uncertain of the relative importance of these sources. The comparatively short residence time for methane in the atmosphere means, however, that there would be a rather rapid return to natural concentration levels if these additional sources were reduced. For further discussion and references see Bolle *et al.* (1986).

Nitrous oxide (N_2O) is another greenhouse gas that is naturally present in the atmosphere (pre-industrial concentrations were about 0.28 ppmv). Its concentration is increasing slowly, the annual increase at present being about

0.3% per year. We estimate that the total increase in comparison with pristine conditions probably has been less than 10%. The natural concentration of nitrous oxide represents a balance between microbial production in soils and surface layers of the oceans and photochemical decomposition in the stratosphere (upper atmosphere). Because transfer to the levels in the stratosphere where decomposition takes place is slow, nitrous oxide has a long average residence time in the atmosphere of about 150 years. The increasing concentration is due to increasing emissions from fossil fuel combustion, biomass burning and the use of artificial fertilisers, which enhance plant growth on land, in lakes and in coastal waters (eutrophication) and thereby increase the generation of nitrogen and nitrous oxide from nitrates in soils (denitrification). We should also take note of the fact that enhanced emissions to the atmosphere cannot easily be stopped, since the present larger amounts of nitrogen in soils and waters due to fertilisation during past decades will decline only slowly and release this excess of fixed nitrogen to the atmosphere for a long time to come. Since the residence time for N_2O in the atmosphere is long, concentrations will continue to increase even if direct emissions by man are reduced and they will not return to pre-industrial levels for several centuries (see, further, Bolle *et al.*, 1986).

The amount of *ozone* in the *lower stratosphere* (20–50 km above ground) is decreasing slowly, by only a fraction of a percent per year (a marked decrease in the Antarctic during a brief spell in spring has recently been detected). It seems most plausible that this ozone decrease is due to the catalytic effect of chlorine which is released when man-made CFCs (chlorofluorocarbons) are decomposed by UV-radiation, although other explanations cannot be altogether excluded. It should be stressed that the drastic decrease of ozone in the stratosphere in the Antarctic was not foreseen, and we must seriously question if the present projection of a slow decrease of ozone in the atmosphere is a valid one. Concentrations in the lower atmosphere (*troposphere*) of the northern hemisphere, on the other hand, are increasing (Fig. 7.10), presumably caused by photochemical reactions induced by air pollutants particularly in the nitrogen oxides, NO and NO_2. This phenomenon has been implicated in recent damage to forests in the northern hemisphere (see Myers, this volume). In addition to these greenhouse gases, which have always been present in the atmosphere, but whose concentrations are now changing, we know that a number of other extraneous greenhouse gases are also being emitted into the atmosphere and staying there for years, decades, and even centuries. In the present context we shall limit ourselves to the consideration of the two most important CFCs, i.e. $CFCl_3$ (F 11) and CF_2Cl_2 (F 12), which probably already play some role in the radiative balance of the earth. The prime concern about their presence in the atmosphere is of course related to their probable catalytic destruction of the ozone layer as briefly referred to above, but we shall be concerned here with their possible climatic effects. The total amount of F 11 in the atmosphere is at present about 6.0 million tons (0.29 ppbv) and the emissions about 0.4 million tons/year. The corresponding figures for F 12 are 8.0 million tons

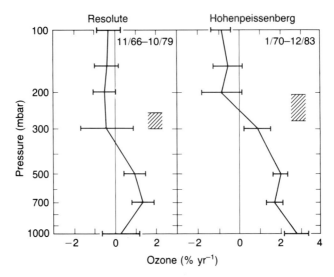

Fig. 7.10. Changes in ozone (in % yr^{-1}) at different heights above ground (expressed in mbar pressure) at Resolute (75°N., Canada, 1966–79) and Hohenpeissenberg (47°N., Germany, 1970–83). The horizontal bars give the 90% confidence interval. The shaded area shows the transition zone between the lower atmosphere (troposphere) and the upper atmosphere (stratosphere). (From Logan, 1985)

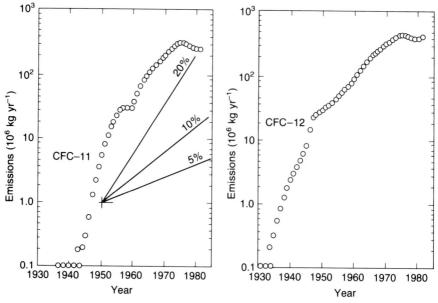

Fig. 7.11. Historical emissions of CFC–11 (CFCl$_3$) and CFC–12 (CF$_2$ Cl$_2$). In the diagram on the left the slopes corresponding to 5%, 10% and 20% annual increase are indicated. (From CMA, 1982)

(0.39 ppbv) and about 0.5 million tons/year. The emissions have been rather constant during the last decade (Fig. 7.11), because of the restriction of their use as spray-can propellants. During the 1980s, however, their use as refrigeration fluids and in plastic foam manufacturing has increased rapidly and the total use is now again rising (see, further, Bolle *et al.*, 1986).

As in the case of N_2O these CFCs, which are in common use, are stable in the troposphere and do not disintegrate until exposed to UV radiation in the stratosphere. Their residence time in the atmosphere is thus about 100 years and they disappear only slowly. Accordingly only a small proportion of the total emissions have so far been decomposed. To prevent their further increase, present emissions must be decreased by 80–90%.

Modification of the earth's radiation balance

When concentrations of greenhouse gases in the atmosphere increase, more of the outgoing IR-radiation from the surface of the earth is absorbed in the troposphere instead of escaping to space. In the case of CO_2, infrared radiation of some wavelengths is already absorbed completely, so increasing the CO_2 concentration further will not affect those wavelengths, but it will increase the range of wavelengths which are absorbed. Similarly increased atmospheric methane and nitrous oxide will absorb an increased range of wavelengths, although their concentrations are much lower and accordingly fewer wavelengths are already absorbed completely. The anthropogenic gases F 11 and F 12, on the other hand, become very efficient since pre-industrial concentrations were zero and also because they absorb wavelengths which are not absorbed by water or carbon dioxide. We can compare the role of these gases in changing the radiative balance of the earth by computing how the atmospheric temperature would change for a given increase of these gases compared to a corresponding increase of atmospheric CO_2. Table 7.1 shows the results of these computations for the idealised case of no clouds and no feedback mechanisms (cf. Ramanathan *et al.*, 1985). It is obvious how very effective the increases of the CFC-gases are. The comparison is quite reliable, since it primarily depends on well-known spectral

Table 7.1. *The relative change of the atmospheric equilibrium temperature due to a given change of volume or mass of the most important greenhouse gases compared to carbon dioxide (based on Ramanathan et al., 1985)*

Greenhouse gas	Relative temperature change due to a given change of:	
	Volume	Mass
CO_2 (carbon dioxide)	1	1
CH_4 (methane)	35	95
N_2O (nitrous oxide)	250	250
$CFCl_3$ (freon 11)	22 000	7000
CF_2Cl_2 (freon 12)	25 000	9000

Table 7.2. *Present (1980) and projected (2030) volume fraction as parts per billion of volume (ppbv) of atmospheric greenhouse gases used as a basis for assessing changes of equilibrium temperature of the atmosphere (based on Ramanathan et al., 1985)*

Gas	1980 (ppbv)	2030 (ppbv)	
		Mean	Range
CO_2	339 000	450 000	
CH_4	1550	2340	1850–3300
N_2O	301	375	350–450
$CFCl_3$ (F11)	0.17	1.1	0.5–2.0
CF_2Cl_2 (F12)	0.28	1.8	0.9–3.5
$CHClF_2$ (F22)	0.06	0.9	0.4–1.9
CH_3CCl_3	0.14	1.5	0.7–3.7
CF_3Cl (F13)	0.007	0.06	0.04–0.1
CF_4	0.07	0.24	0.2–0.3
O_3 (troposphere)		+12.5%	
O_3 (stratosphere)			
10 km		+3.8%	
22 km		+4.5%	
26 km		+2.0%	
30 km		−6.1%	
34 km		−22.6%	
40 km		−37.9%	
50 km		−5.5%	

data and only marginally on the climate model which is used. Table 7.1 can be used to express increases of the other greenhouse gases in terms of an approximately equivalent increase of CO_2.

The estimated temperature rises due to greenhouse gases may need to be increased further by 50–100% to account for positive feedback, because a general heating of the atmosphere will result in an increase of water vapour in the atmosphere and this in turn will absorb more infra-red.

Since we do not know how the concentrations of these greenhouse gases will change in the future we are not able to forecast, in the strict sense of the word, future changes of climate. Instead, the important thing is to illustrate what possible climatic changes might occur as a result of different assumptions of future concentrations. Ramanathan *et al.* (1985) have designed a few such scenarios and computed the associated changes of the equilibrium temperature, taking into account the absorption wavelengths for the different gases. Table 7.2 shows the assumptions made and Fig. 7.12 shows the estimated resulting changes of the global surface temperature due to CO_2 and other atmospheric gases. Since a simplified model of the atmosphere considers only the vertical structure of the atmosphere, the values should be increased by 50–100% to account for likely feedback mechanisms. Neverthe-

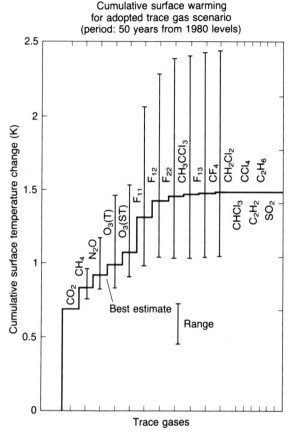

Fig. 7.12. Cumulative equilibrium surface temperature warming due to increase in CO_2 and other trace gases, from 1980 to 2030. (After Ramanathan *et al.*, 1985)

less, the *relative* magnitudes of the changes induced by the different gaseous constituents shown in Fig. 7.12 are presumably reasonably accurate.

In conclusion the computations show, that in 1980 the estimated change of the equilibrium temperature was $\frac{1}{3}$ due to the increase of greenhouse gases other than CO_2 and $\frac{2}{3}$ due to the CO_2 increase. Further, the mean values projected for 2030, as given in Table 7.2, would yield a further increase of the mean equilibrium temperature at the earth's surface due to the other greenhouse gases even larger than that due to the assumed increase of the atmospheric CO_2 concentration to 450 ppmv. The increase of CO_2 and additional greenhouse gases as projected in Table 7.2 would together be equivalent to about a doubling of the pre-industrial atmospheric CO_2 concentration by only 2030. The possible future climatic change due to the combined effect of these man-induced emissions into the atmosphere may become one of the most important environmental problems during the first decades of the next century.

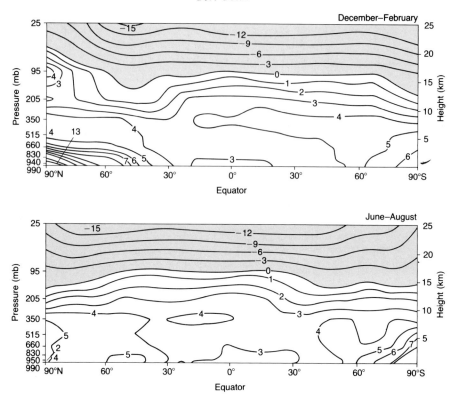

Fig. 7.13. Average temperature changes in °C versus altitude and latitude for quadrupling of CO_2 emissions. Stippled areas indicate decreased temperatures. (After Manabe & Stouffer, 1980)

How will climate change?

It is difficult to assess the likely regional distribution of climatic changes on earth induced by the change of the energy absorption of the atmosphere discussed above. It must be based on elaborate models of the interactions between the atmosphere and the oceans and also of the role of land surface processes, particularly induced changes of the hydrological cycle (including snow and ice) and terrestrial ecosystems. Although a number of models have been developed with somewhat different treatment of the relevant physical processes, the assessments of the likely *regional* distribution of changes are not yet reliable.

There is general agreement that a doubling of the atmospheric CO_2 concentration (or the equivalent CO_2 concentration, including the effects of increasing concentrations of other greenhouse gases) would increase the mean equilibrium temperature at the earth's surface between 1.5 and 4.5°C (cf. Dickinson, 1986). The temperature in the lower stratosphere, on the other hand, would decrease (Fig. 7.13). It should be emphasised, however, that the oceans would delay the expected warming of the Earth's surface by decades. Model computations suggest that the mean global warming resulting from increases in greenhouse gases to date might be in the range 0.3°C–1.1°C (Bolle *et al.*, 1986). The actual increase during this period (the last 120 years) has been 0.5 ± 0.2°C (cf. Fig. 7.3). Still, we cannot conclusively say that this observed increase is primarily due to the increasing concentrations of greenhouse gases. Other changes of our global environment are taking place. Our knowledge about their role in changing the climate is still limited.

Fig. 7.13 shows that the polar regions (depicted at the ends of the horizontal axes) are more affected if a general warming occurs. The amount of sea ice would decrease, which in turn would change the albedo of the polar regions. This positive feedback is actually included in the model computations shown in Fig. 7.13. In a longer time perspective the major ice sheets would also change, but their response time is of the order of centuries to millennia. The Greenland ice sheet would probably decrease, but enhanced precipitation might well increase rather than decrease the accumulation in the Antarctic. Still, a warming of this magnitude is expected to lead to a sea level rise between 25 and 165 cm, primarily because of the thermal expansion of the upper layers of the ocean (Robin, 1986).

We are not able to derive much further detail about the spatial distribution of a global warming induced by increasing concentrations of greenhouse gases, except that there is some evidence that mid-latitude, mid-continental drying in the summer might result. It does not seem likely that we shall be able to deduce more conclusive results before model computations can be compared with an ongoing change that has been clearly established by global observations. That can possibly be done before the end of the century.

The difficulty of detecting an *ongoing* climatic change is primarily due to the spatial and temporal variability of the natural climate. For this reason we are not able to tell if the repeated droughts during the last decades in the Sahel area in Africa, and similarly the eastern parts of Brazil, are extreme variations around an unchanging climate or signs of a long-term trend towards a drier climate.

Terrestrial ecosystems will undoubtedly be affected by these projected changes of climate. At high latitudes where moisture will be adequate, the cold marginal areas will retreat polewards. Forests and possibly agriculture will expand and the rate of growth will be enhanced. On the other hand, the semi-arid and arid regions, particularly in the sub-tropics, may shift significantly in location and change in extent. Countries located in such marginal regions with a deteriorating climate might be hit severely (cf. Warrick *et al.*, 1986). We will only learn gradually where and how people and their settlements on earth may be influenced, nor do we know when changes may become significant for nations and for the global community. Possibly the climatic system is more robust than our models tell us. If this is the case then the changes will be less drastic, but this can hardly be used as a justification for ignoring the true possibility of future marked climatic changes.

It is of course not meaningful to analyse possible changes beyond the next century. Interesting questions arise, however. We recall, for example, that the two polar regions have been ice covered during the last several million years. The characteristic response time of the Antarctic ice-sheet certainly is thousands of years, although the west Antarctic ice-sheet might possibly disintegrate gradually in a matter of centuries in the case of sustained warming. The sea-ice in the North Polar Sea, on the other hand, is only about 3 m thick on average and might disappear in summertime during the next century, if a sustained warming were established. This in turn would represent a drastic change of the albedo and the re-establishment of the sea-ice in winter might be hampered. It was pointed out at the begining that we are presumably on our way towards another ice-age within a few millennia, if man were not interfering with the natural course of events. The question arises if a warmer climate induced by man might prevent this happening. Certainly, we do not know, but it seems more likely that however disruptive a major change of climate due to increased concentrations of atmospheric gases might be for people and nations on earth, it probably would be modest and of short duration seen in the geological perspective. The characteristic 'Milankovic quasi-periodic' variations of climate on earth take place on a timescale of about 20 000 to 100 000 years. It seems intuitively plausible that these would return, when the disruptive episode came to an end.

Further reading

Arrhenius, S. (1896). On the influence of carbonic acid in the air upon the temperature of the ground. *Phil.Mag.*, **41**, 237.

Adel, A. (1939). Note on the atmospheric oxides of nitrogen. *Astrophys.J.*, **90**, 627.

Bacastow, R. & Keeling, C.D. (1981). Atmospheric carbon dioxide concentration and the observed air-born fraction. In *Carbon Cycle Modelling*, SCOPE 16, ed. B. Bolin. Chichester: Wiley, pp. 103–12.

Bolin, B. (1986). How much CO_2 will remain in the atmosphere? In *The Greenhouse Effect, Climatic Change and Ecosystems, SCOPE 29* eds, Bolin, B., Döös, B., Warrick, R. & Jäger, J. Chichester: Wiley, pp. 93–155.

Bolle, H.J., Seiler, W. & Bolin, B. (1986). Other greenhouse gases and aerosols. In *The Greenhouse Effect, Climatic Change and Ecosystems*, SCOPE 29, eds. Bolin, B., Döös, B., Warrick, R. & Jäger, J. Chichester: Wiley, pp. 157–203.

Callender, G.S. (1938). The artificial production of carbon dioxide and its influence on temperature. *Q.J.R.Met. Soc.*, **64**, 223.

Dickinson, R.E. (1986). The climate system and modelling of future climate. In *The Greenhouse Effect, Climatic Change and Ecosystems*, SCOPE 29, eds. Bolin, B., Döös, B., Warrick, R. & Jäger, J. Chichester: Wiley, pp. 207–70.

Douglas, R.C. & Hughes, T.J. (1981). Deep sea benthic foraminifera. In *The Sea*, vol. 7, ed. Emiliani, C. New York: Wiley.

Jones, P.D., Wigley, T.M.L. & Wright, P.B. (1986). Global temperature variations between 1861 and 1984. *Nature*, **322**, 430–4.

Keepin, W., Mintzer, I. & Kristoferson, L. (1986). Emissions of CO_2 into the atmosphere. In *The Greenhouse Effect, Climatic Change and Ecosystems*, SCOPE 29, eds. Bolin, B., Döös, B., Warrick, R. & Jäger, J. Chichester: Wiley, pp. 35–91.

Lovelock, J. (1979). *Gaia: A new look at life on earth*. Oxford University Press.

Neftel, A., Moor, E., Oeschger, H. & Stauffer, B. (1985). The increase of atmospheric CO_2 in the last two centuries. Evidence from polar ice cores. *Nature*, **315**, 45–47.

Ramanathan, V., Cicerone, R.J., Singh, H.B. & Kiehl, J.T. (1985). Trace gas trends and their potential role in climatic change. *J. Geophys. Res.*, **90**, D 3, 5547–66.

Robin, G. de Q. (1986). Changing the sea level. In *The Greenhouse Effect, Climatic Change and Ecosystems*, SCOPE 29, eds. Bolin, B., Döös, B., Warrick, R. & Jäger, J. Chichester: Wiley, pp. 323–59.

Rotty, R. (1987). A look at 1983 CO_2 emissions from fossil fuels (with preliminary data for 1984). *Tellus*, **39B**, 203–8.

Rotty, R. & Marland, G. (1986). Fossil fuel combustion; recent amounts, patterns and trends of CO_2. In *The Changing Carbon Cycle; a Global Analysis*, eds. Trabalka, J. & Reichle, D. Springer Verlag, pp. 484–500.

Shackleton, N.J. & Pisias, N.G. (1985). Atmospheric carbon dioxide, orbital forcing and climate. In *The Carbon Cycle and Atmospheric CO_2: Natural Variations, Archean to Present*, eds. Sundqvist, E.T. & Broecker, W.S. Geophys. Monogr. 32, pp. 303–18. Washington DC: Am. Geophys. Union.

Siegenthaler, U. and Oeschger, H., (1987). Biospheric CO_2 emissions during the past 200 years reconstructed by deconvolution of ice core data. *Tellus* 39B, 140–54.

Warrick, R.A., Shugart, H.H., Antonovsky, M.Ya., Tarrant, J.R. & Tucker, C.J. (1986). The effects of increased CO_2 and climatic change on terrestrial ecosystems. In *The Greenhouse Effect, Climatic Change and Ecosystems*, SCOPE 29, eds. Bolin, B., Döös, B., Warrick, R. & Jäger, J. Chichester: Wiley, pp. 363–92.

[8]

Observing earth's environment from space

Gordon Wells

Stories told by distant travellers have long provoked the imaginations of placebound listeners. One such account telling about the adventures of a young former Cambridge student during his five-year circumnavigation of the earth was published in 1839. In *The Voyage of the Beagle*, Charles Darwin gives us his first narrative exploration of experiences and reflections that eventually created a lasting revolution in the sciences.

Today we live in an age when travellers can circle the earth in approximately 90 minutes. Their stories and photographs, when coupled with the data returned by unmanned satellite sensor systems, can be the guide to new discoveries about our environment. In this chapter I propose to take the reader on an orbital photographic journey around the earth and to demonstrate that many of the issues discussed by the other contributors to this series can be examined through evidence seen by unaided vision from the altitude of the Space Shuttle orbiters, averaging about 280 km. Since in no way can one attempt to do justice to all NASA programmes of terrestrial observation and the many investigations currently pursued by other agencies and university research groups, I shall choose to focus on several topics of personal interest. These include the fate of tropical forests, the nature of the African drought environment, the factors changing global climate, the geological hazards posed by particular volcanoes and the dynamics and biology of the world's oceans.

Exploring the earth from orbit

Before beginning to consider environmental topics, it is useful to recall the historical development of earth surveillance from space. Early experiments with V-2, Aerobee and Viking rockets launched from the White Sands Proving Ground in New Mexico during the late 1940s and early 1950s recorded panoramic images of the American southwest from altitudes of 100–250 km. However, these first examples of what were quaintly referred to as 'hyperaltitude photographs' sparked little interest until NASA astronaut

Gordon Cooper returned from the final orbital mission of the Mercury series in May 1963 with spectacular photographs made with a hand-held Hasselblad camera of regions remote from western view. In one classic example, his photographs of central Tibet showed terrain that had not been seen by western geoscientists since the remarkable explorations of Swedish geologist Sven Hedin at the turn of the century and the even more improbable 1895 traverse of the central plateau by St George Littledale, his wife, his nephew of Oxford rowing fame and the family's spirited fox terrier. That photographs made from an orbital altitude could supply information about the geology, vegetation and hydrology of such an inaccessible area confirmed the scientific value of exploring the earth from space.

During the same period of the 1960s, instruments carried by the early Television Infrared Observation Satellites (TIROS) began to relay continental views of global cloud patterns. In 1966, NASA launched the first meteorological satellite to be placed into a geostationary orbit for the continual hemispheric monitoring of weather phenomena important to daily forecasts and tropical cyclone warnings. Concurrent with the development of meteorological satellites were the launches of the Ranger, Surveyor and Lunar Orbiter probes to the Moon. The camera systems of these spacecraft relayed detailed images of the lunar surface in order that site selections could be made for the Apollo manned missions. In a short time the robotic imaging technologies that previewed the lunar landings were turned to face the earth.

The Landsat satellite series is the result of lunar programme technology brought back to earth. During the 1970s, these polar-orbiting platforms provided the first detailed, multispectral images of virtually the entire land surface of our planet. The data returned by Landsat sensors opened new research horizons in the geosciences, including the tectonic studies of the Alpine and Himalayan mountain chains, performed by Dan McKenzie of the Cambridge Department of Earth Sciences.

The decade of the 1980s has been highlighted by a multinational effort in orbital earth observation. NASA has launched advanced versions of the Landsat satellite and carried a variety of imaging radars, multi-spectral scanners, film cameras and other terrestrial sensor systems in the payload bays of the Space Shuttle orbiters. The French CNES has received superb images from its SPOT satellite, while the Japanese NASDA is now collecting images from its recently launched marine observation satellite, MOS-1. By the close of the decade, the European Space Agency will orbit the ERS-1, its first satellite constructed for detailed reconnaissance of the ocean and land surface using radar systems with a large component of British design development by Marconi Space Systems.

With so much recent emphasis upon the images returned from unmanned satellites, it is important to distinguish the differences between these sensor system images and the photographs made by Space Shuttle astronauts. A Landsat earth observation satellite circles the planet in a sun-synchronous, polar orbit such that it crosses regions at approximately the same local time throughout the year and is capable of imaging the same area every 16 days.

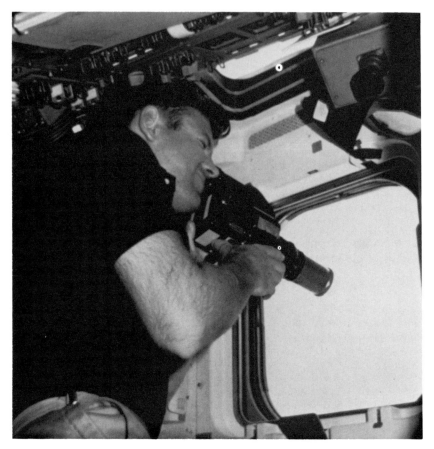

Fig. 8.1. Astronaut Michael Coats with Hasselblad camera (NASA Nikon).

For Landsat, the sun elevation above the horizon changes marginally according to the seasons. The electronic sensors of the Landsat Thematic Mapper view regions directly below the satellite and record both visible radiation and portions of the reflected and thermal infrared spectrum not detected by human vision. When the satellite controllers instruct the sensor system to collect data, the images of earth are sent in the form of digital telemetry to receiving stations for processing. Though the final Thematic Mapper image may resemble a conventional photograph, it is composed of many individual picture elements each about 30 m by 30 m, with every full Landsat TM scene containing slightly more than 38 million pixels.

By contrast, Space Shuttle astronauts take photographs of earth using Hasselblad and Linhof camera systems exposing natural colour film in most instances. Their views through the orbiter windows permit a range of perspectives from a direct nadir survey to sweeping oblique panoramas. The orbits of a Space Shuttle mission create a constantly changing sun elevation across the surface that furnishes dramatic illumination for cloud formations

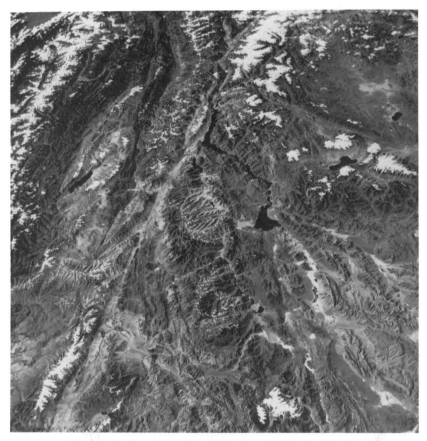

Fig. 8.2. Karakoram Fault, western Tibet (NASA Hasselblad).

and mountain ranges, when the sun is low on the horizon. In order to prepare for photography, an astronaut may check a windowside guide for the observation sites that will be encountered during a particular orbit. The guide will briefly state the significance of sites discussed during crew training and will mention appropriate techniques for photography. An astronaut's primary information will come from a Shuttle Portable Onboard Computer, which Space Shuttle crews fondly call Mr SPOC. The computer screen shows a world orbit map much the same as the one in the Mission Control Operations Room in Houston. The screen displays the instantaneous position of the orbiter, the locations of upcoming observation sites and the proper camera settings adjusted for local sun elevation and terrain brightness for photographs made over the planned sites or of areas of unexpected interest below the spacecraft. By following these procedures, the astronauts return with perfectly exposed colour photographs conveying the best record of how the earth appears to an observer in space.

We can come to appreciate the wide-ranging diversity of global landscapes when seeing the astronauts' photographs. Across the western Tibetan plateau (see Fig. 8.36) one encounters a region of high, arid beauty with almost no evidence of human presence. The environment of a few thousand Ladaki herdsmen, their goats and yaks is representative of perhaps 40% of the world's land surface yet to be marked by human activities. These regions are found in the Antarctic continent, the boreal forests and permafrost wildernesses of North America and Siberia, the high plateaus of the Andean Altiplano and Tibet, the interior reaches of tropical forests in the Amazon Basin and equatorial Africa and in the hyperarid deserts of north Africa, southwest Asia and central Eurasia.

Across the southern High Plains, my own state of Texas offers an example of the opposing extreme in the many cultural patterns one sees upon the landscape. Over a region that was once the prairie grassland of western cattle drives, an orbital view displays a quilt of geometric field patterns, where cotton, grain sorghum and wheat are grown, and circular plots, where the rotating booms of centre-pivot irrigation systems water crops of maize and

Fig. 8.3. West Texas High Plains agriculture (NASA Hasselblad).

alfalfa. This being Texas, it comes as no surprise to find on closer inspection a multitude of white dots indicating the placement of oil field wellheads. In the case of the southern High Plains, the entire landscape has been converted to human purposes. Perhaps 60% of the global land surface demonstrates some degree of large-scale conversion.

Even in the most hostile environments for human occupation, orbital images disclose the efforts of mankind to alter the natural landscape through the use of modern technology. An early Landsat scene recorded in 1972 of the central An Nafud Desert in Saudi Arabia reveals rock outcrops, barren plains and large, mobile sand dunes. A map of the area made at the time traces a single caravan route across the image towards the border with Iraq. A dozen years later, a NASA astronaut photographed the same area now containing clusters of green circles, the result of centre-pivot irrigation systems tapping a fossil water aquifer. The new artificial oases of the An Nafud testify to the measures a culture will take to modify the natural landscape. Though considering a history of living with the natural extremes of the desert, one can appreciate Saudi enthusiasm for an ability to make the sterile sands green; the sudden appearance of these changes, their scale and prospects for future development raise many questions about our understanding of their long-term consequences. To determine the rate at which anthropogenic land surface alteration is occurring within the global environment is a principal goal of surveillance using data from satellite sensor systems and orbital photography.

The status of tropical forests

Tropical forests are both the richest biotic environments in the numbers of plant and animal species and the most fragile landscapes in terms of the biological pressures brought about by the rapid clearing of woodland for timber, fuel-wood and agriculture (see Myers, this volume). One of the most striking characteristics of orbital photographs made over the Amazon Basin is the preferential convective cloud development above the forest, while river courses in the same region remain relatively clear. What would first be taken as a surface temperature difference, with the warmer forest promoting greater convection, turns out to be a quite different phenomenon. In most cases, thermal measurements made from satellites indicate little temperature difference between the vegetated land and the water surfaces. The distinctive cloud patterns are created by the literal breathing of the vegetation. In a dense area of moisture-saturated, multi-tiered rainforest, leaf surfaces may respire water vapour in amounts greater than can be evaporated from a nearby open surface of water. One can readily appreciate the prospects for climatic change, if the tropical rainforest is removed from a region.

Each year during the last weeks of July through September, fires are set to consume crop residue remaining in the recently cleared fields of the Amazon Basin. The daily images from the National Oceanic and Atmospheric Administration (NOAA) Polar Orbiter satellites first reveal a few, isolated

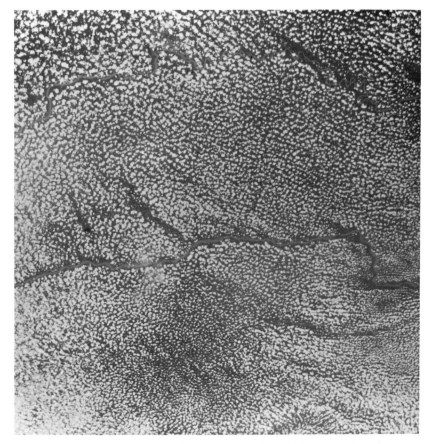

Fig. 8.4. Rio Meta cloud patterns, Colombia (NASA Hasselblad).

smoke palls. A day or two later several fires may be concentrated in one region. At the peak of the burning season, the smoke obscures the entire basin, and regions the size of Wales may have hundreds of fires burning at one time with cumulus clouds forming atop the most intense thermal plumes.

Interest in the atmospheric chemistry of seasonal Amazon burning was stimulated by the detection of a large carbon monoxide anomaly over northern South America by the Measurement of Air Pollution from Satellites (MAPS) experiment carried during the second Space Shuttle mission in November 1982. Field research in Brazil coordinated with aircraft and balloon surveys has been directed by a group from the NASA Langley Research Center. Their work in the summer of 1985 for the Amazon Boundary Layer Experiment (ABLE) helped to define the processes responsible for vertical mixing of atmospheric trace molecules emitted by the rainforest and of the carbon monoxide, carbon dioxide and various hydrocarbons released by burning vegetation. One discovery of the project is the doubling of carbon monoxide in the Amazon boundary layer toward the conclusion of the

Fig. 8.5. Rondonia agricultural fires, Brazil (NASA Hasselblad).

burning season, along with levels three to ten times higher than earlier measurements in the haze layers observed at higher altitudes.

Much of the slash-and-burn agriculture practised by farmers in the Amazon Basin takes place in new, state-sponsored projects promoting the movement of peasants from overpopulated coastal areas into the interior of Brazil. The largest colonisation projects are found in the state of Rondonia in south-western Brazil, where baseroads penetrate the tropical rainforest to form a settlement grid for 100-hectare farm plots. Peasant farmers raise subsistence crops, such as maize, using slash-and-burn methods with little use of fertilisers, crop rotation or regard for soil properties. As the initial arable land becomes exhausted, additional forest is cleared for cultivation.

In order to quantify the scale and rate of land conversion in the Amazon Basin, satellite images collected by the NOAA Polar Orbiters and Landsat can be processed to distinguish the density of vegetation. Both satellites employ sensor systems that record visible and reflected infrared radiation. The chlorophyll in the leaves of green vegetation absorbs almost all blue and red light, while reflecting a moderate amount of green light. To our eyes, the vegetation will appear green. In the portion of the spectrum immediately beyond red light, the reflected infrared, plant chlorophyll is almost transparent. Much infrared radiation is reflected by the internal structures of leaves, making them appear very bright to a sensor measuring the reflected infrared. Conversely, bare soils reflect large amounts of visible radiation, but less of the infrared than does green vegetation. By taking advantage of these physical properties, an image-processing technique that determines the ratio

Fig. 8.6. Rondonia baseroad and field grid, Brazil (NASA Hasselblad).

Fig. 8.7. Pan American highway initial baseroads, Brazil (NASA Hasselblad).

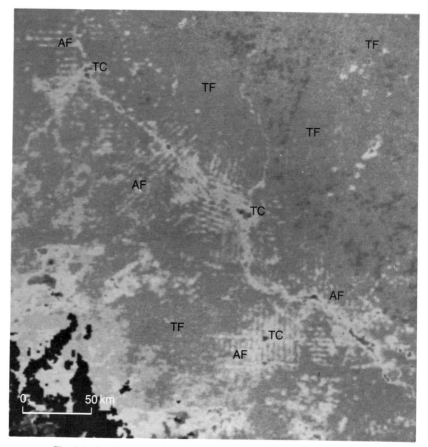

Fig. 8.8. Rondonia normalised difference vegetation index, Brazil.
Tropical forest (TF); agricultural fields (AF); township centres (TC)
(NOAA AVHRR).

of reflected infrared radiation to visible radiation for each picture element will discriminate the relative abundance of vegetation within areas of a region.

A method to study Amazon Basin deforestation by deriving vegetation indices from NOAA Polar Orbiter data has been applied by Chris Justice of the University of Maryland, Brent Holben and Jim Tucker of NASA Goddard Space Flight Center and John Townshend of the University of Reading. Though the spatial resolution of the data is rather low, with each picture element measuring slightly more than one square kilometre, it is adequate to detect the large-scale colonisation projects in Rondonia and has the advantage of providing coverage of broad regions at low costs. To supplement regional analyses with Polar Orbiter data, George Woodwell of the Woods Hole Research Center in Massachusetts and his colleagues have processed a series of Landsat images of Rondonia for dates beginning in the early 1970s. The greater spatial resolution of Landsat data increases the precision of classification programmes determining areas of tropical forest clearance.

Fig. 8.9. Rondonia land clearance 1976–1981 (Landsat MSS).

The published estimates for the rate of deforestation occurring in the Amazon rainforest vary by an order of magnitude. When considering the entire region, many of these predictions are excessively high. In the instance of Rondonia, over the period since the early 1960s, the annual rate of land conversion has been about 0.2%. Yet the rate may be accelerating in the areas adjacent to the developing network of baseroads. Though across

Fig. 8.10. Betroka tropical deforestation, Madagascar (NASA Hasselblad).

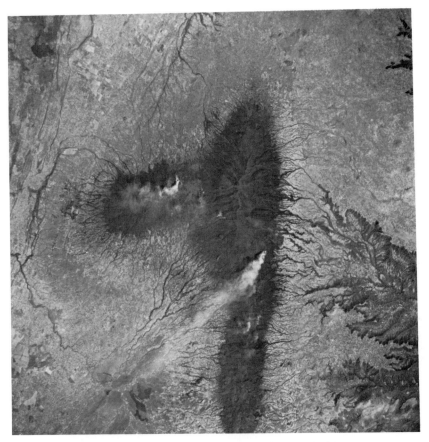

Fig. 8.11. Amher mountains deforestation, Ethiopia (NASA Hasselblad).

the entire Amazon Basin, primary tropical forest currently may be consumed at an annual rate of 0.2–0.5%, once growing populations become established in the interior along with the better integration of transportation systems, the rate of land clearance may increase exponentially. The results of preliminary investigations argue strongly for a satellite program monitoring the region to identify the absolute scale of land surface alteration and the areas of accelerating land conversion.

The consequences of failure to place limits on the clearing of tropical forests are best exemplified by a study of astronaut photographs of the island of Madagascar. Often Space Shuttle astronauts have returned from missions remarking about the 'bleeding island' with its rivers emptying red sediment into the Mozambique Channel and Indian Ocean. The forests of the northwestern portion of Madagascar have been decimated to expose easily eroded soils and underlying soft geological units. Botemboka Bay, an estuary of the Betsiboka River, was once navigable by sizable ships, but has recently filled with the silt of a rapidly aggrading delta. Upstream tributaries of the Betsiboka dissect a landscape of rills and alcoves removing colossal amounts of sediment, as much as 250 metric tons of soil per hectare each year in one area.

Over the past decades, Madagascar has lost more than four-fifths of its rainforest to land clearance for cattle rangeland, cropland, timber and fuel-wood. To my mind, one of the saddest photographs ever returned from space shows the region near Betroka in the southeastern part of the island. Here isolated refuges of rainforest appear in the midst of a nearly denuded exposure of metamorphic rock. One has great sympathy for the lemurs studied by Cambridge's Alison Jolly, when seeing the reduction of their habitat to these fragments. Perhaps better than any statistical summary or recitation of erosion rates, a photograph of this kind delivers the message of the extent of environmental degradation. In view of the changes apparent from orbit, the combined fate of the island's tropical forests and the approximately 120 000 species of animals and plants found only on Madagascar may lie in the immediate prohibition of future forest removal and a national programme of rapid afforestation.

The loss of woodland cover in the tropics should not be considered solely a problem of rainforest clearance. In many underdeveloped, semi-arid regions, the cutting of forest and brushland for fuel-wood has increased with the expanding population and rising cost of fuel from petroleum products. The acacia brushland areas of the western Sahel have village barrens appearing from orbit as rough circles enclosing communities. The radius of each circle in many areas represents a day's journey and return on foot with a limited supply of wood for cooking. In the population centres of Ethiopia and Kenya, large amounts of charcoal are imported from neighbouring regions. The remaining areas of dense woodland in the two nations are under severe pressure from local charcoal production. Throughout the semi-arid regions of northeastern Africa and across the Sahel, the problem of woodland depletion has been compounded by persisting drought.

Monitoring the African drought environments

For the period from 1968 to the present, rainfall has been below the historical mean recorded from 1931–60 during almost every year in each region of the western Sahel from central Chad to coastal Senegal and Mauritania. The environmental repercussions of the long drought have staggered the resource base of Sahelian nations, which are now largely dependent upon western aid. In tracing the changes to the landscape, the orbital photography of NASA astronauts has been particularly useful because views recorded during the Gemini missions of the mid-1960s predate available satellite images of sub-Saharan Africa and show the environment prior to the onset of drought. The extent of Lake Chad, a body of water in a basin without external outlet, is a good indicator of the severity of climatic deterioration. Photographs from the Gemini 9 mission in June 1966 show a lake expanse of about 22 000 km^2. In the lake were many islands formed by the waters surrounding ancient sand

Fig. 8.12. Inland delta of the Niger, Timbuktu, Mali, in 1965, with grassland extending well to the north of Timbuktu (T) (NASA Hasselblad).

Fig. 8.13. Prolonged drought in the inland delta of the Niger, leading
to loss of vegetation cover and low lake levels by 1985 (NASA Hasselblad).

dunes. An economy supported by fishing and cereal production flourished
around the unique lake environment, where the sand dune islands were
named for their villages.

By the summer of 1985, evaporation of the lake had long since led to the
collapse of these cultural patterns. Photographs from the Space Shuttle
orbiters at that time defined an open water area of only 2500 km^2. Though
substantial rainfall over the southern and eastern regions supplying Lake
Chad led to its doubling in area to 5000 km^2 by January 1986, the lake failed to
reach a level allowing water to restore large-scale irrigation schemes, such as
the Nigerian South Chad Irrigation Project, a £200 million investment
designed to be the centrepiece of regional development but predicated upon
the lake levels of 20 years ago. The failure of the project since 1983 has forced
thousands of native farmers to leave their land to raise crops and graze cattle
near the retreating shoreline upon the lake floor in areas covered by metres of
water only a few years ago.

In many areas of the western Sahel, the regional spread of desertification can be best appreciated by reviewing past and recent orbital photography. During the Gemini 6 mission in December 1965, Walter Schirra and Thomas Stafford made photographs looking over the celebrated town of Timbuktu in central Mali. These views indicate dense vegetation in the Niger River floodplain and within the many distributary channels of the Inland Delta, where the river cuts across ancient sand dunes. To the north of Timbuktu, open grassland extends for 150–200 km approaching the mobile sands of the Sahara. Twenty years later in April 1985, a Space Shuttle astronaut photographed the same area from a similar perspective. The floodplain vegetation had diminished to a small fraction of its 1965 extent, and lakes in the Inland Delta were entirely dry except for two small basins near the main river course. Most disturbing of all the observed effects, every sign of open grassland to the north of the Niger River had vanished. The removal of savanna grassland fringing the Sahara during the drought period can be traced in a swathe of similar width from western Chad to the southern Mauritanian coast. In the Sahelian regions to the east, the evidence for desertification processes having altered the landscape is much more sporadic.

Orbital observations of the river channels crossing the Sahel can contribute information about the changing status of irrigation projects and their relative merits in meeting the needs of the region. Astronaut photography excels in this activity through its capacity to record oblique photographs of the specular reflection of sunlight striking open surfaces of water. The technique sharply outlines the level of water in river courses and even allows the irrigation conduits and secondary trellis patterns to be observed. Over a period of several years, the photography can document the success or failure of vast irrigation schemes, such as the South Chad Irrigation Project, and enable the comparative study of many smaller-scale projects that may demonstrate a more flexible response to drought conditions.

Global satellite television transmissions of field reports by BBC correspondents and those of other networks during the summer of 1984 brought home the severity of drought and famine conditions in Ethiopia, Sudan and the western Sahel. However, long before these broadcasts alerted the general public, earth observation satellites provided data that clearly predicted the worsening situation. For the authorities of the UN Food and Agricultural Organization and national relief administrators, the problems created by intense drought and accompanying food shortages are amplified by the remoteness of many affected areas, the lack of transportation, the necessarily subjective nature of surface field reports and the difficulty in collecting timely, comprehensive regional analyses.

A monitoring program for Sahelian drought environments has been in operation for several years at the NASA Goddard Space Flight Center guided by the work of Tucker, Holben, Justice and Townshend. Weekly vegetation indices for the Sahel are composed from Polar Orbiter data sets representing the collection of images during several orbital passes over the region. Each composite vegetation index displays the highest reflected infrared to visible

Fig. 8.14. Niono Irrigation Project, Niger River, Mali (NASA Hasselblad).

Fig. 8.15. Namibia/Angola border desertification (NASA Hasselblad).

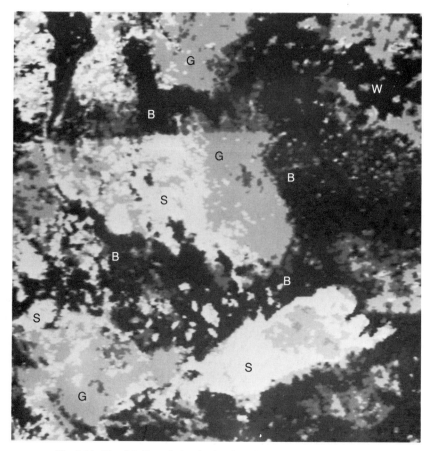

Fig. 8.16. Namibia/Angola border land surface cover. Exposed soils (S); grassland (G); brushland (B); deciduous woodland (W) (NOAA AVHRR).

light ratio recorded during the observation period for each of the co-registered picture elements. From the beginning of the monsoon rains in early summer to the northernmost penetration of convective rainfall in August and September, the weekly vegetation indices record the relative abundance of healthy vegetation. During a year with adequate, well-timed rainfall, the images will show progressive greening of Sahelian grassland to the Saharan perimeter. Now after several years of the program, interannual comparisons can be made on an area-by-area basis for each stage of the summer precipitation season. Knowledge of area crop yields and pasture conditions from field investigations conducted during previous years permits analysts to predict coming shortfalls in time to recommend where to send limited food reserves. In this manner, relief coordinators can obtain a more objective view of the absolute scale of drought effects, so that food shipments can be procured and field logistics planned. The areas most in need of assistance can be selected for immediate relief efforts.

Photographs made by Space Shuttle astronauts also contribute to the use of satellite images to monitor desertification. When the timing of astronaut photography closely coincides with satellite data acquisition over a region, the photographs can be used as a critique of atmospheric conditions for the presence of dust, smoke and haze. Only the data collected during periods of unusual atmospheric clarity may be suited for detailed studies of surface features. One outstanding example of coordinated photography and satellite imaging comes from a series of views exposed by the late Judith Resnick during her first Space Shuttle mission in 1984. Her photographs made in the early afternoon of 1 September reveal the abrupt national boundary between Angola and Namibia, a feature produced by differing land use practices on either side of a barbed wire fence. The NOAA-8 Polar Orbiter had collected data over southwest Africa several hours earlier that morning. Confidence in the absence of atmospheric turbidity on that day led Lockheed scientist Robert Mohler working at NASA Johnson Space Center to apply a surface cover classification technique to the image developed by Lockheed's Robert Cate. In this procedure, visible, reflected infrared and thermal infrared measurements are combined to determine types of land cover. The results are displayed in various colour codes to identify areas of grassland, exposed soils, acacia brushland and deciduous woodland. A series of surface cover classification images made during the next several years may show changes in the desertification features found in the region.

The truism that national boundaries cannot be seen from space is false. The example of the land cover differences along the border between Angola and Namibia also focusses attention on the multicausal nature of land degradation processes. Northern Namibia is the wettest region of the territory, while southern Angola is the driest part of that nation. More intensive land use would be anticipated to occur on the Namibian side of the border for this reason alone. But the circumstances promoting desertification are exacerbated by the apportioning of the region as a native homeland, Ovamboland, where the growing population is constrained to live upon the overtaxed land. One might add that border crossings by South Africans into Angola may serve as an additional disincentive for Angolan agriculturalists to occupy the southern border region. In this case, climatic zones, drought, overpopulation and political policies can be seen acting together to affect the landscape.

Earth's changing climate – past and future

The effects of currently changing climatic patterns are most noticeable in regions undergoing catastrophic droughts. Yet signs of much greater long-term excursions of climate are preserved in the landscapes of many regions of the world. Even within the Amazon Basin there are clues to past episodes of desertification and sand desert formation. An orbital photograph of the Santa Cruz region in central Bolivia displays the reactivated sand dunes of an

Fig. 8.17. Santa Cruz ancient sand dunes (SD), Bolivia (NASA Hasselblad).

ancient dunefield recently covered by tropical forest until woodland clearance exposed the underlying desert landforms. During the final stages of the last Ice Age, between about 18 000 and 12 000 years ago, the sand deserts of the earth were much larger and regions of modern rich farmland were created by the deposition of windblown dust along the perimeters of the retreating ice sheets. As Dick Grove of Downing College, Cambridge, has confirmed during his many explorations of sub-Saharan Africa, the landscape of the Sahel with its ancient sand dunes and indurated soils was created at the time of deglaciation.

A comparison of maps locating modern sand deserts and those that existed during the waning stages of the last Ice Age corroborates the magnitude of climatic change in recent earth history and underscores the vulnerability of particular regions to drought. In the Sahel, Kalahari, High Plains of North America, the Argentine pampas, western Siberian steppes, Indus Valley plain and the semi-arid areas of Australia, the landscape and soils were formed

[170]

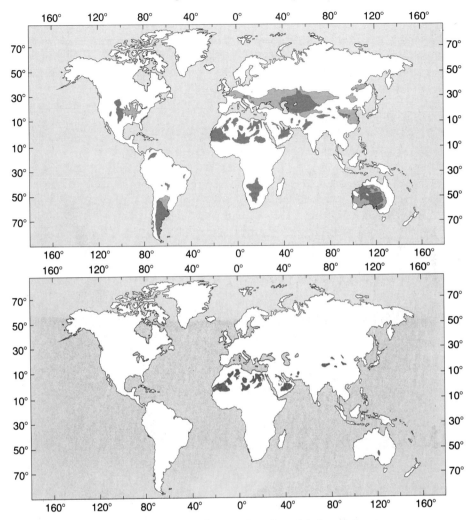

Fig. 8.18. Ice Age (above) and modern global sand deserts.

during periods of intense aridity at the conclusion of the Ice Age. Each region is predisposed to desertification during drought by the nature of its soil and landforms. Each region has been a modern drought environment during this century. When one considers that these areas produce the great majority of all the surplus food in the world, then one realises the importance of understanding the relationship between the processes that formed the landscapes and those that in the future may transform them, when assisted by man's alteration of the environment to meet his needs.

Nowhere are the lessons in more need of application than in the High Plains of the USA, the world's principal supplier of cereal grains and livestock products. A rapid expansion of agriculture into marginal lands through the use of centre-pivot irrigation technology occurred during the early 1970s.

[171]

Today, 6% of the state of Nebraska is a centre-pivot field. In virtually every case, these irrigation systems are employed to farm areas of ancient sand dunes once anchored by native grassland. The destabilisation of these dunefields, coupled with the rapid depletion of the Ogallala aquifer providing regional irrigation water, sets the stage for a return to the conditions of 50 years ago should a prolonged drought begin. Though major dust storms are currently infrequent events in the southern High Plains, the exposure of vast areas of vulnerable land invites the inception of a future dust bowl that will mirror the extremes now recalled as distant memories.

To appreciate the scale of dust storms sweeping across the Sahel requires the viewpoint achieved from orbital altitudes. Space Shuttle astronauts have photographed palls of dust completely obscuring the surface of the Atlantic Ocean over 2400 km from the west African coast. Enormous bow wakes are generated as the dust passes around the Cape Verde Islands and Sao Tiago, where the *Beagle* and Charles Darwin landed in 1832. Darwin made the first

Fig. 8.19. Tropical convection over Texas coast (NASA Hasselblad).

scientific observations of African dust transport during this period of persistent Sahelian drought, and published his results in the *Quarterly Journal of the Geological Society of London* in 1846. The source areas of the Harmattan dust storms lie deep in the western Sahel and southern Sahara of Chad, Niger and Mali. The transit of dust from these areas leads to mineral aerosol transport across the ocean basin, as proved by dust collection over Barbados and other regions of the western Atlantic.

I have an interest in the influence of dust upon convective rainfall. In my area of coastal Texas, the summer climate is dominated by the rains generated from large thunderstorms. In these rain-making systems, we see cumulus clouds rising through the clear atmosphere as a consequence of the daily surface heating promoting convective turbulence. The ascent of these clouds resembles that of a hot-air balloon. Under these circumstances, clouds can continually expand their surface areas through the condensation of water vapour and the release of latent heat. So long as air surrounding the cumulus

Fig. 8.20. Dust-inhibited convection over Mauritania (NASA Hasselblad).

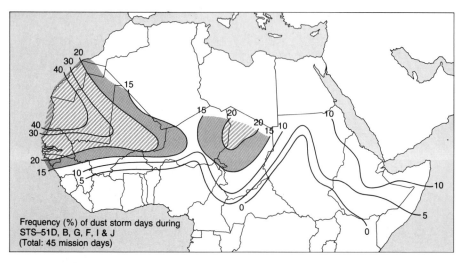

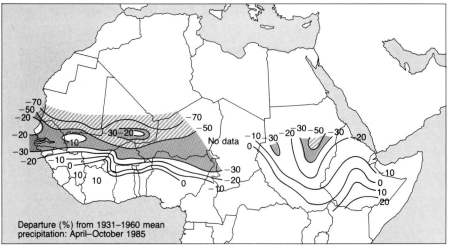

Fig. 8.21. Sahelian dust storm days (top), documented photographically during Space Shuttle missions July–October, 1985, and precipitation (below), recorded at surface meteorological stations.

cloud is cooler than that inside, it will continue to rise buoyantly and perhaps mature as a giant thunderhead that reaches the base of the stratosphere, where the air temperature becomes isothermal.

Astronaut photographs of cloud formations embedded in dust storms over the western Sahel indicate the operation of a distinctly different process. Though incipient cumulus clouds are formed near the surface, as they rise into the dust veil, the clouds become increasingly stratoform. Few cumulus clouds penetrate the top of a dust pall, and almost none ascend to the altitudes necessary for ice nucleation that leads to the production of raindrops. Measurements by instruments carried aloft in balloons demon-

[174]

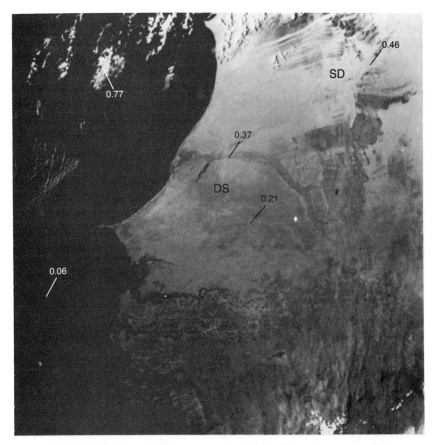

Fig. 8.22. Dust storm (DS) over western Sahel reflects almost as much incoming radiation as areas of sand dune (SD) (NOAA AVHRR).

strate that dust veils form strong temperature inversions, as the dust creates a thermal blanket trapping the infrared radiation emitted by the land surface during the daily heating cycle. The dramatic change in the otherwise uniform cooling of the atmosphere with increasing altitude serves to inhibit rainfall.

In an attempt to quantify the degree to which dust suppresses the incidence of convective precipitation across the Sahel, I mapped the distribution of dust storms observed during the six Space Shuttle missions of April through October 1985. These monthly missions provided daily observations for more than one-quarter of the entire period. The rainfall distribution during the summer monsoon season of 1985 led to a break of drought conditions in the eastern Sahel and Ethiopia and to a continuation of dry conditions across the western Sahel. A comparison of the number of dust storm days in a region and the surface meteorological station precipitation reports for the period shows a striking correlation between the presence of a dusty atmosphere and limited rainfall.

The frequent occurrence of dust storms may limit regional rainfall in another manner. Several years ago, Jules Charney proposed that the continuing Sahelian drought could be the result of land surface alteration. According to his general circulation model, an increase in the brightness (albedo) of the Sahelian surface to a degree equal to the albedo of the Saharan sand deserts would promote regional subsidence of dry air masses overlying the Sahel. These descending air masses could block the intrusion of monsoonal moisture during the summer rainfall season. Recent analyses of satellite imagery of the Sahel conclude that a wholesale brightening of the land surface has not occurred. On the other hand, albedo measurements of dust storms, sand deserts and the Sahelian land surface using data from the NOAA Polar Orbiters show that the typical dust veil reflects almost as much incoming radiation as does bare sand. In a revision of the Charney hypothesis, the albedo change created by frequent dense dust palls in the lower troposphere may create regional subsidence of the dry air masses over the Sahel.

As Nick Middleton of the Oxford University School of Geography has reported, the frequency of dust storms in the western Sahel has increased seven-fold since the beginning of the drought in 1968. Though dust storms have long been regarded as symptomatic of prolonged drought, the photographs taken from the Space Shuttle reveal that dust may be the *cause* of extended drought. At both the level of individual cloud dynamics and the regional scale of air mass circulation, the dust appears capable of sustaining drought conditions. If the dust mechanisms suppressing rainfall are the primary causes of the continuing Sahelian drought, then the measures required to return the climate to more favourable conditions will involve restoration of the disturbed land surface in large areas of recent denudation disclosed by the record of orbital photography.

The changing chemistry of the global atmosphere has attracted much scientific scrutiny, as current research seeks to resolve the effects of increasing levels of carbon dioxide, chlorofluorocarbons and hydrocarbons. The unusually low levels of ozone detected over the Antarctic continent during the past decade by the Total Ozone Mapping Spectrometer carried onboard the Nimbus-7 satellite have sparked a debate on the influence of ozone-depleting trace molecules and seasonal changes of Antarctic circulation. Many of the issues explored by such investigations could be better studied if a more comprehensive data set existed for the relative abundance, vertical mixing characteristics and global distribution of the major trace molecule species. An experiment designed by Barnard Farmer and his research team at the NASA Jet Propulsion Laboratory gathered this kind of information during a Space Shuttle mission in May 1985. Future flights of the Atmospheric Trace Molecules Observation by Spectroscopy (ATMOS) experiment will help to quantify the residence times, sources and sinks and chemical interactions of more than 20 atmospheric constituents.

The climatic effects of the El Chichon eruption in 1982 impressed upon the geoscientific community the importance of sulphur-rich volcanic eruptions in changing global climate. Though El Chichon is a rather insignificant volcano by all other standards, the unusual chemistry of its eruption products led to a decline in northern hemisphere temperatures lasting several months, when gaseous sulphur dioxide injected into the stratosphere combined with water vapour to form droplets of sulphuric acid which reflected incident solar radiation. The photographs and visual observations of Space Shuttle astronauts have illustrated the problems of monitoring volcanic eruptions. On 13 April 1981, during the first Space Shuttle mission, a photograph was made of an eruption plume from Lewatola volcano on Lomblen Island in the Indonesian archipelago. No surface report or other documentation of the eruption was received by western scientists. On 31 August 1983, the same region of Indonesia was photographed by an astronaut. On the adjacent island of Andonara, Ili Boleng volcano was erupting. Again, no other evidence exists for the eruption. Perhaps the inhabitants of these islands expect their volcanoes to erupt and see no reason to send out special notice of

Fig. 8.23. Ili Boleng volcano eruption, Andonara Island, Indonesia (NASA Hasselblad).

renewed activity. In many cases, orbital photography has provided the initial information about an erupting volcano, while in several instances, it supplies the only record. If the dynamic chemistry of the global atmosphere is to be determined with precision, orbital monitoring techniques need to be applied for the detection of all small-scale and remote volcanic eruptions.

Collapsing volcanoes

The cataclysmic eruption of Mount St Helens on 18 May 1980 gave a dramatic demonstration of the geological hazards present in the environment. The event was initiated by an earthquake that caused the collapse of an unstable volcanic dome and triggered a powerful lateral blast destroying most of the northern sector of the composite cone and spreading a debris avalanche across the landscape. Such eruptive behaviour was first believed to be peculiar for a volcano. A detailed review of studies on the geological settings and eruptive histories of global volcanoes conducted by Lee Siebert of the Smithsonian Institution found 73 other cases where an eruption was accompanied by the destruction of a sector of the volcanic cone. Intrigued by this evidence, Peter Francis of the Lunar and Planetary Institute and Open University began a search of the volcanoes on the central Andean Altiplano looking for the results of past volcanic eruptions similar to the Mount St Helens episode. For two years, we have been working together to analyse Landsat Thematic Mapper data and Space Shuttle astronaut photography of the Central Andes. The dry climate and high altitude of the region make it an ideal locale for observation from space and for the preservation of geological features created by volcanoes with sector collapses.

Our survey has led to the identification of 42 volcanoes between 16°S and 28°S that have undergone sector collapses, with 14 of these displaying well-preserved debris avalanches. One of the most spectacular examples is located in the Nevados de Payachata area of northernmost Chile, where the collapse of the western sector of Parinacota volcano produced a debris avalanche that dammed a local river to create Lake Chungara and provided the setting for Lake Cotocotani, a famous wildlife refuge that has formed in the middle of the Parinacota avalanche deposit. Radiocarbon dating of lacustrine peats has allowed us to estimate that the eruption occurred shortly before 13 500 years ago. Some of the most memorable scenes from the recent BBC *Flight of the Condor* series were filmed along the shores of Lake Cotocotani without the realisation that the terrain owed its origin to the prehistoric collapse of Parinacota. The volatile character of Parinacota is difficult to appreciate, when one views the beautifully symmetrical modern composite cone. Since the retreat of glaciers in the region, eruptions of ash and lava have entirely resurfaced the volcano to cover all traces of the scars left by its sudden sector collapse.

It is difficult to convey the violence of events involved in the creation of the now serene landscape surrounding Lake Cotocotani. In the case of a debris

Fig. 8.24. Prehistoric debris avalanche (DA) on Parinacota volcano, and Lake Cotocotani (LC), Chile (NASA Hasselblad).

Fig. 8.25. Parinacota volcano and Lake Cotocotani, Chile (P.W. Francis photo).

avalanche from a volcano to the south of Parinacota, our calculation for the flow dynamics yields a minimum velocity of 270 km h^{-1}. In a matter of minutes, a landscape can be formed that will last for many hundred thousand years. There appear to be at least three factors that lead some volcanoes on the central Andean Altiplano to collapse. One is the extremely arid climate that limits erosion and permits the construction of extremely steep slope profiles. While most volcanoes in the world have summit slope angles of 30° or less, those in the central Andes often have summit slopes exceeding 35°. These oversteepened volcanoes are gravitationally disposed to a sudden collapse. Another factor is the relatively high viscosity of lavas in the Central Andean volcanic province. The andesites and dacites of the region tend to form short, thick lava flows extending from summit vents. The third contributing influence seems to be local tectonic activity. A large proportion of the sector collapses have taken place in quadrants oriented perpendicular to regional fault trends. Vertical movement along normal faults would tend to

Fig. 8.26. El Misti volcano and Arequipa (A), Peru (NASA Hasselblad).

pitch a destabilised, oversteepened volcanic cone towards one of these quadrants.

A study of the causes of debris avalanches in the central Andes could be seen as an entirely academic pursuit, if it were not for the presence of suspect volcanoes near large population centres. Arequipa, a Peruvian city of almost one million inhabitants, lies in a valley oriented perpendicular to the regional fault trend and 14 km from the summit of El Misti, an oversteepened composite volcano. Though El Misti has not erupted during historical times and only occasionally vents sulphuric steam, the absence of glacial features on its slopes indicates that major eruptions have occurred during the past 12 000 years. As is the case with Mount Rainier in Washington and Mount Fuji in Japan, even a large *nuée ardente* eruption, such as swept down Mount Pelée and through St Pierre in 1902, would cause a major disaster. A sector collapse of such a volcano could produce one of the greatest calamities in human history.

Oceanography from space

To an orbiting observer, the Hawaiian Islands appear a tranquil chain, as cloud streets following the trade winds form wakes in their passage around the islands. Each of the volcanic islands has an abrupt sub-surface shelf, where the gently sloping layers of lava composing the shield volcanoes above surface produce steep flow fronts below sea level because lava does not move easily through water. James Moore of the United States Geological Survey has reported enormous landslides created by the failure of these oversteepened slopes. In one instance approximately 100 000 years ago, a tsunami generated by such a landslide swept over the island of Lanai at a height greater than 300 m.

Much more subtle phenomena can be detected by observation from the Space Shuttle orbiters. During the El Niño event of 1982–83, astronaut photographs of the Hawaiian Islands made as sunlight reflected from the ocean surface surrounding the islands revealed long wakes in the waters leeward of each island. Shipborne measurements at that time in the Kaiwi Channel separating Oahu and Molokai confirmed a $1.0–1.5 \text{ m s}^{-1}$ acceleration in the velocity of the North Equatorial Current. According to some accounts, Polynesian mariners can navigate toward islands beyond the horizon by determining the sea surface texture and current differences in the waters extending downstream from open channels and island platforms. The initial detection of an El Niño event often involves noting changes in the sea surface temperatures of areas in the Peru Current imaged by the thermal infrared radiometers on board the NOAA Polar Orbiter satellites. When the upwelling of cool waters relaxes in the region off coastal Peru, sea surface temperatures may rise by 2.0–3.0°C, as warm equatorial water spreads across the area of diminished upwelling.

The sea surface temperature anomalies that appear during an El Niño are connected to a system of geophysical feedback mechanisms that link atmospheric pressure, trade wind patterns and ocean currents across the entire Pacific basin. When an extremely strong El Niño, such as the 1982–83 event, reaches its maximum intensity, the waters of the Peru Current may warm by as much as 6.0°C, the regional trade winds may reverse direction and equatorial ocean current velocities may markedly increase. Recent climatic research suggests that these conditions are responsible for a variety of changing weather patterns leading to drought in Australia and the eastern Pacific and unusually high rainfall along the Atacama coast of Peru and northern Chile.

In the future, the ocean-observing sensors of the European Space Agency's ERS-1 satellite will allow earlier identification of the onset of El Niño conditions. The ERS-1 Wind Scatterometer will relay radar reflectivity measurements of the ocean surface to be processed to determine wave orientations and to infer wind velocity and direction. When these calculations are coupled with the sea surface temperature images collected by the ERS-1 and other satellites, the wind pattern and water temperature changes indicative of the early stages of an El Niño event should become apparent.

Several of the most exciting earth science discoveries made with photography from the Space Shuttle orbiters have come in the realm of ocean dynamics. For more than a decade, sea surface temperature measurements from satellites have allowed oceanographers to study the evolution of large-scale cold- and warm-core eddies that develop along major ocean current

Fig. 8.27. Strait of Gibraltar internal waves (NASA Hasselblad).

boundaries, such as the Gulf Stream. But between the megascale region of these large features having diameters of more than 100 km and the point-sampling perspective provided by shipborne surveys, oceanographers have gained limited knowledge of the mesoscale dynamics of oceanic circulation. Orbital photographs that record the mirror reflection of sunlight off the ocean surface reveal subtle differences in sea surface texture produced by the orientation of capillary waves. These 'sun glint' views permit the detection of circulation features that would go unnoticed by visual observations from on board a ship.

A spectacular example of the mesoscale dynamics of oceanic circulation was photographed by oceanographer Paul Scully-Power during his Space Shuttle mission in October 1984. The volume of water evaporated from the Mediterranean Sea creates a constant influx of Atlantic water through the Strait of Gibraltar. Inflowing Atlantic water is accelerated by passage through the narrow strait, while the denser, more saline Mediterranean water sinks beneath it. At the interface between the two water masses, located at a depth of 60–80 m, internal waves are generated. Sets of internal waves are pulsed through the strait with the changing tide of the region. Though the surface expression of the internal wave packets, photographed by astronaut Scully-Power, makes them appear enormous, their actual surface amplitude was only about 5 cm, as measured by instruments on board a research vessel transiting the region during the time of photography.

As Atlantic waters flow into the Mediterranean Sea, they come into contact with Mediterranean currents having a different water salinity, density, temperature, velocity and direction. Where these water masses collide, shear zones are formed. Spiral eddies are often created along the wall of a shear zone. These cyclonic systems having diameters of 10–20 km were first detected in photographs made from the Space Shuttle orbiters. Their widespread occurrence in the global oceans indicates that spiral eddies are a fundamental feature of mesoscale circulation. Though models of ocean dynamics are only now beginning to incorporate such features, they appear to embody the turbulent, mathematically non-linear processes responsible for the translation of oceanic energy, while the internal wave systems represent the more laminar, mathematically linear processes.

In many ocean regions, the natural oils exuded by phytoplankton enhance the presence of dynamic structures by highlighting the current streamlines seen in the sun glint. The abundance of ocean biology can also be observed directly. Following the ninth Space Shuttle mission in December 1983, Commander John Young and Pilot Brewster Shaw were curious to learn the nature of a mass of green filaments that they had photographed in the southern Capricorn Channel separating Queensland from the Great Barrier Reef. By chance, a NOAA Polar Orbiter satellite collected data over the area twelve minutes after the astronauts' photography. Analysis of the satellite image revealed that the green mass spread along 600 km of the coastal lagoon and was as bright in the reflected infrared as vegetated areas on the coast of Queensland. This led to the conclusion that the mystery

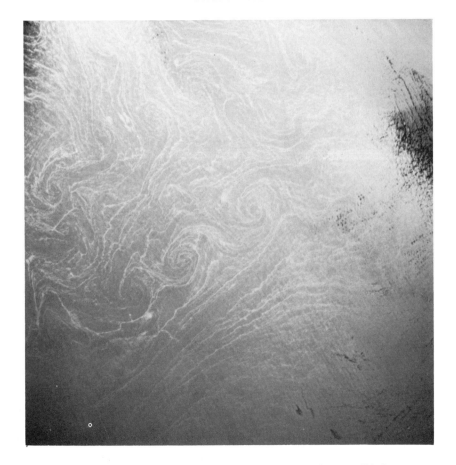

Fig. 8.28. Mediterranean shear wall spiral eddies (NASA Hasselblad).

substance was actively photosynthetic. Further investigation by Deborah Kuchler and Nigel Arnold of the Australian CSIRO Division of Water Resources Research subsequently confirmed from surface reports that the green mass was a giant bloom of the blue–green 'alga' *Oscillatoria erythraea*. The green filaments detected by the astronauts and satellite sensors were a massive phytoplankton bloom swept down the Capricorn Channel by longshore currents.

Astronauts often observe the blue–green accumulations of phytoplankton in areas of oceanic upwelling, the pink masses of zooplankton that feed upon the plant species and, in the cold waters of the southern Atlantic, the scarlet concentrations of krill that consume plankton. A counterpoint to their witnessing elements of the food chain and distribution of ocean biology is the astronauts' documentation of ocean pollution. The most prevalent polluting activity is the bilging of water ballast from supertankers. When not carrying a cargo of 5–7 million gallons of crude oil, a supertanker is somewhat fragile. Water ballast is taken into the oil storage tanks to ensure that the ship keeps a

Fig. 8.29. Supertanker bilging off Benghazi, Mediterranean (NASA Hasselblad).

low profile in the water. Upon approaching a crude oil loading facility, supertankers pump out the ballast and with it a residue of petroleum. The oil-coated stern wakes trailed by bilging supertankers have been traced for over 100 km in photographs made by astronauts. Whereas the natural oils of phytoplankton form an almost monomolecular layer across the sea surface, the paraffin component of heavy crude petroleum creates slicks of coagulated oil that are easily discerned in the ocean sun glint. Though observations from the Space Shuttle orbiters have alerted oceanographers to the common practice of supertanker bilging in the open oceans, a programme to monitor routinely the pollution of the world's seaways has yet to be initiated.

Darwin and the space programme

I have no intention of ending this volume by leaving the reader with an unfortunate vision of man's abuse of the ocean environment, but would prefer to leave you with a more paradisical view enjoyed by Charles Darwin and described in *The Voyage of the Beagle*. On the morning of 17 November 1835, Darwin landed with the *Beagle* at Matavai Bay on the shore of western Tahiti. After breakfast, he climbed the volcanic slopes of a nearby foothill to an altitude of about 700 m. There he looked out across the sea to the island of Moorea 25 km in the distance. From his vantage point, he saw the cloud-capped peaks of the green volcanic island outlined against the blue sky in a blue Pacific broken only by the white line of waves crashing against an

Fig. 8.30. Tahiti and Moorea, Society Islands (NASA Hasselblad).

encircling coral reef. The view made a deep impression upon Darwin, who compared the sight to 'a framed engraving, where the frame represents the breakers, the marginal paper the smooth lagoon, and the drawing the island itself'. In the evening, he retraced his steps to the harbour, ate a dinner of hot roasted bananas and a pineapple, drank the milk of a young coconut and began to reflect upon what he had seen.

Within several weeks, he had developed the essential details of a theory that crystallised around his experience looking out from the hillside above Matavai Bay. From his observations of Pacific coral reefs that form fringing reefs, barrier reefs and atolls with shallow lagoons, he inferred an evolutionary sequence. When a volcanic island such as Tahiti or Moorea, is young, coral growth begins close to the shoreline and creates a fringing reef. As the island becomes older, it sinks progressively into the sea, while the corals continue to grow upward from their original encircling location. A barrier reef develops separated from the eroded volcanic island by a wide lagoon. With continuing subsidence, the ancient island massif disappears beneath the surface to leave a coral atoll sheltering a broad, central lagoon. Though Darwin could not offer an exact explanation for the process causing volcanic islands to sink gradually beneath the sea, his theory demonstrating the stages of coral reef development remains the core of our understanding more than 150 years after his visit to Tahiti.

It pays to climb mountainsides in search of a better view. In our age, this includes the climbing of a gravitational mountain to a height of 280 km. Had Darwin the chance to survey the earth from the altitude attained by the Space Shuttle orbiters, he could have seen his theory of coral reef evolution

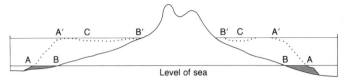

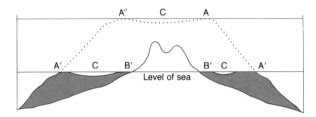

Fig. 8.31. Darwin's coral reef theory. The island slowly sinks (sea levels A, A' and A") while the coral continues to grow, from a fringing to a barrier reef (top) to an atoll (below) (engraving from *The Voyage of the Beagle*).

illustrated by a single scene in its own way as striking as his hillside picture of Moorea. Moving west from the younger volcanic islands of Tahiti and Moorea in the Society Island chain, one encounters the successively older islands of the Iles sous les Vents. When volcanic activity ceased, the magma chambers supporting these island platforms began to subside into the crust. The islands of Hauhine, Raiatea and Tahaa retain barrier reefs with lagoons that broaden according to the age of the island and depth of its submergence. Only

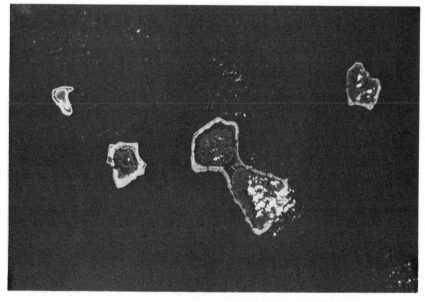

Fig. 8.32. Iles sous les Vents, Society Islands (NASA Hasselblad).

fragments of the once lofty volcanoes of the still older Bora-Bora remain above water, while the most ancient volcanic island of the chain has completely vanished, leaving atop its descending platform the atoll of Motu Iti.

I want to ask the reader to consider what might have been the course of Darwin's thought had he forgone the opportunity to sail on the *Beagle* and, instead, had arranged with Captain FitzRoy to receive reports, sketches and samples collected during the global voyage for Darwin to study at home in England. Even in the miraculous coincidence that the same specimens were to be collected and phenomena noted, I have no doubt that the theoretical conclusions reached by Darwin would have been quite different. No one can read *The Voyage of the Beagle* without appreciating the connection between Darwin's experiences making his observations and the ultimate expression of these original impressions in his theoretical work. His insight was guided by direct experience, and it would be difficult to separate exactly which experiences were necessary to frame his later scientific inquiry. Certainly, the climbing of the foothill above Matavai Bay proved to be important, as did witnessing the earthquake at Valdivia and crossing the Andes on muleback and collecting fossil seashells at 4000 m. But one can imagine incidental observations made while riding with the gauchos across the Pampas or while enduring passages through the Strait of Magellan also contributing to the spirit of his later investigations.

When I hear questions about the role of manned spaceflight in exploration of the earth from space, an analogy to Darwin's experience and scientific

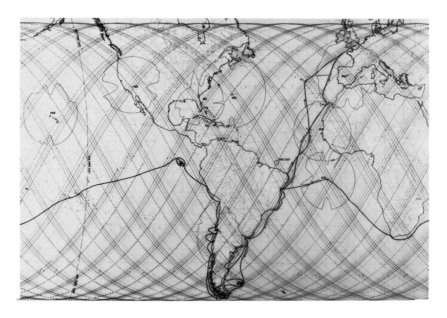

Fig. 8.33. Voyage of the *Beagle*, duration five years, and STS–41G orbital groundtracks for five days of travel aboard a Space Shuttle orbiter.

Fig. 8.34. NASA space station (design illustration).

contributions seems particularly appropriate. For over two decades, unmanned earth observation satellites have accumulated data leading to hundreds of discoveries in the geosciences. These satellites perform their operations for a fraction of the cost of manned spacecraft and therefore should arguably receive primary support for future earth science missions to be conducted from orbit. Yet the limitations of these robotic remote sensing systems need to be acknowledged. They function best when gathering data collected under a narrow range of constraints, including instrument viewing direction, spectral range, spatial resolution, sun elevation and direction of illumination. Their instruments are often preprogrammed to obtain measurements hours or days in advance, and offer little opportunity for interaction with ground controllers, when data are actually collected. In short, remote sensing satellites are inflexible systems useful to collect repetitive sets of data under precisely predetermined conditions. With this in mind, one should not be surprised to learn that many initial discoveries of new phenomena, such as those recently reported in ocean dynamics, have been made by astronaut observers. Though their records are made with relatively simple film cameras, the astronauts possess the ability to contemplate and adjust their observations while watching features that capture their interest. The inherent flexibility of human observers to detect, track and study new terrestrial phenomena needs to receive broader recognition in the scientific community.

By far the best location for pursuing manned exploration of the earth from orbit will be from an earth observatory on a future space station. From such a viewpoint, an orbiting geoscientist could coordinate his investigations with research relying upon information returned from satellite remote sensing instruments. He could preview his approaching observation opportunities by studying the most recent images from polar orbiting and geostationary satellites made available to him by telemetry transmitted from ground receiving stations and relay satellites. In turn, he could alert ground controllers to the location and nature of significant environmental features, so that satellite images could be obtained. From the space station, he could document his observations with conventional cameras and with mobile sensor pallets attached to the station super-structure that he guides by remote control. If these instruments were to take the form of my design for a Pointable Interactive Camera System (PICS), then earth-based scientists could share in the moment of discovery by watching high-resolution television images transmitted directly to the mission control centre. During those times when the orbiting observatory was unoccupied, the pointable imaging system could be operated by ground personnel who conduct observations by sighting with the television camera.

With several weeks to adapt physically to the microgravity environment, an earth scientist on a space station could begin to assimilate the experiences offered by his unique perspective on the planet. Just as Darwin came to realise the subtle interconnectedness of natural phenomena viewed on a global scale, the orbiting scientist might gain insight into earth processes beyond the scope of current inquiry. Very likely he would prove again in

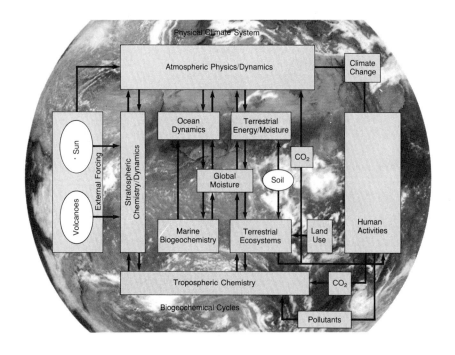

Fig. 8.35. Earth systems science cycles and flows.

his explorations that the power of our intuition exceeds that of our instrumentality.

In the future, the central aim of orbital geoscience research will be to discover and quantify the various geophysical cycles interlinking the atmosphere, biosphere, oceans and solid earth. The new earth systems science will attempt to model the interactions of these complex units on all timescales with the intention of detecting alterations occurring within each component and predicting the consequences to the system as a whole. To study global change created by natural processes or human activities requires the collection of carefully registered sets of data generated by repeated measurements. A variety of orbital electronic remote sensing instruments will be required to construct such a data base. The role of manned observation from spacecraft will be to discover the missing links and unknown parts of the global biogeophysical system. If we can come to an understanding of the interrelationship of earth's environments and can numerically specify their interactions, then we will have the ability to predict their future course and the effects of human intervention of natural cycles.

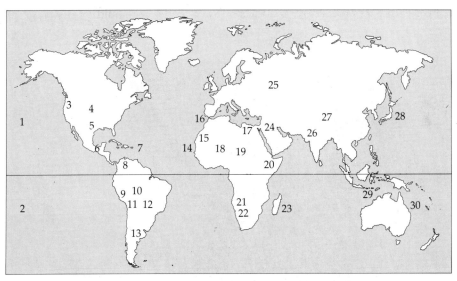

1 Hawaiian Islands	11 Central Andes, N Chile	21 Angola/Namibia border
2 Society Islands, Tahiti	12 Santa Cruz, Bolivia	22 Kalahari
3 Mt St Helens, Mt Rainier	13 Argentine Pampas	23 Madagascar
4 USA High Plains	14 Cape Verde Islands	24 An Nafud
5 Texas coast	15 Mauritania	25 W Siberian Steppes
6 El Chichon	16 Strait of Gibraltar	26 Indus Valley
7 Mt. Pelée, Martinique	17 Mediterranean coast, Benghazi	27 Western Tibet
8 Rio Meta, Colombia	18 Inland Delta, Timbuktu, Mali	28 Mt Fuji
9 Arequipa, El Misti, Peru	19 Lake Chad	29 Ili Boleng, Andonara
10 Rondonia, Brazil	20 Amher Mts, Ethiopia	30 Capricorn Channel

Fig. 8.36.

Towards the conclusion of the Apollo lunar landing programme, the American editor and poet Norman Cousins wrote: 'On the way to the Moon, we discovered the Earth.' Looking back to the planet from space, we continue to discover the earth. The future depends upon the understanding gained from a global overview of earth's environments, or we face the prospect that through our negligence we will reduce the world to parts that can no longer be integrated. I believe we are lucky. At the same time our technology permits global ecological change, we have developed the means to monitor its impact and plan for our future with greater foresight.

Further reading

Allen, J.P. & Martin, R. (1985). *Entering Space*. New York: Stewart, Tabori & Chang.

Francis, P.W. & Jones, P. (1984). *Images of Earth*. London: George Philip & Son.

Francis, P. & Self, S. (1987). Collapsing volcanoes. *Scientific American*, **256** (6), 34–40.

Francis, P.W. & Wells, G.L. (1988). Landsat Thematic Mapper observations of debris avalanche deposits in the Central Andes. *Bulletin of Volcanology*, **50**, 258–78.

Glantz, M.H. (1987). Drought in Africa. *Scientific American*, **256** (6), 90–7.

Holz, R.K. (1985). *The Surveillant Science*, 2nd edn. New York: John Wiley & Sons.

Justice, C.O. (ed.) (1986). Monitoring the grasslands of semi-arid Africa using NOAA AVHRR data. *International Journal of Remote Sensing* (special issue), **7**, 1383–1622.

La Violette, P.E., Yentsch, C.S. & Apel, R. (1987). The oceanographer in space: the next step. *EOS 68*, **121**, 130–1.

Lillesand, T.M. & Kiefer, R.W. (1979). *Remote Sensing and Image Interpretation*. New York: John Wiley & Sons.

Wiener, J. (1985). *Planet Earth*. New York: Bantam.

NASA photo numbers

Fig. 8.1. NASA S14–1512	*Fig. 8.2.* NASA S17–31–046
Fig. 8.3. NASA 61A–41–080	*Fig. 8.4.* NASA S07–19–0864
Fig. 8.5. NASA S14–40–021	*Fig. 8.6.* NASA 51G–34–061
Fig. 8.7. NASA 51I–33–069	*Fig. 8.10.* NASA S14–35–086
Fig. 8.11. NASA 41B–41–2405	*Fig. 8.12.* NASA S65–63248
Fig. 8.13. NASA S85–23–32–033	*Fig. 8.14.* NASA 61B–32–039
Fig. 8.15. NASA 51G–46–078	*Fig. 8.17.* NASA 51I–40–071
Fig. 8.19. NASA 51F–37–076	*Fig. 8.20.* NASA 51G–47–047
Fig. 8.23. NASA S08–50–1840	*Fig. 8.24.* NASA S19–45–053
Fig. 8.26. NASA S17–35–021	*Fig. 8.27.* NASA S17–34–081
Fig. 8.28. NASA S17–35–094	*Fig. 8.29.* NASA S17–38–063
Fig. 8.30. NASA S08–46–0886	*Fig. 8.32.* NASA S08–46–0881

Information concerning the purchase of Space Shuttle photography may be obtained from EROS Data Center, User Services Section, Sioux Falls, South Dakota 57198.

Index